T0325715

Data Treatment in Environmental Sciences

"To avoid interpreting what we do not understand, we need to understand methods and details before making sense of the conclusions"

Statistiques, méfiez-vous! (Gauvrit, 2014)

Series Editor
Françoise Gaill

Data Treatment in Environmental Sciences

Multivaried Approach

Valérie David

First published 2017 in Great Britain and the United States by ISTE Press Ltd and Elsevier Ltd

Apart from any fair dealing for the purposes of research or private study, or criticism or review, as permitted under the Copyright, Designs and Patents Act 1988, this publication may only be reproduced, stored or transmitted, in any form or by any means, with the prior permission in writing of the publishers, or in the case of reprographic reproduction in accordance with the terms and licenses issued by the CLA. Enquiries concerning reproduction outside these terms should be sent to the publishers at the undermentioned address:

ISTE Press Ltd
27-37 St George's Road
London SW19 4EU
UK

www.iste.co.uk

Elsevier Ltd
The Boulevard, Langford Lane
Kidlington, Oxford, OX5 1GB
UK

www.elsevier.com

Notices
Knowledge and best practice in this field are constantly changing. As new research and experience broaden our understanding, changes in research methods, professional practices, or medical treatment may become necessary.

Practitioners and researchers must always rely on their own experience and knowledge in evaluating and using any information, methods, compounds, or experiments described herein. In using such information or methods they should be mindful of their own safety and the safety of others, including parties for whom they have a professional responsibility.

To the fullest extent of the law, neither the Publisher nor the authors, contributors, or editors, assume any liability for any injury and/or damage to persons or property as a matter of products liability, negligence or otherwise, or from any use or operation of any methods, products, instructions, or ideas contained in the material herein.

For information on all our publications visit our website at http://store.elsevier.com/

© ISTE Press Ltd 2017
The rights of Valérie David to be identified as the author of this work have been asserted by her in accordance with the Copyright, Designs and Patents Act 1988.

British Library Cataloguing-in-Publication Data
A CIP record for this book is available from the British Library
Library of Congress Cataloging in Publication Data
A catalog record for this book is available from the Library of Congress
ISBN 978-1-78548-239-7

Printed and bound in the UK and US

Contents

Introduction

I.1. Why is numerical analysis used in environmental science?

I.1.1. Researchers handling their data

"Environmental" research often involves sampling in the field or the automated acquisition of several parameters (i.e. chemical–physical, pedological, biological) in different stations and/or on different dates. Each station (or date) represents an **observation/object** that will be, therefore, defined by several abiotic or biotic **descriptors/variables/parameters**.

The variation of one or two of these descriptors is easy to assess with a simple graphical analysis. For example, it is possible (1) to observe the monthly fluctuations of river discharges over a seasonal cycle at one estuarine station, (2) to determine which dates show similar discharges and are therefore similar concerning this parameter (for example flood and low-water periods) or (3) to analyze whether there is a relationship between river discharges and another parameter such as turbidity (Figure I.1(A)).

However, researchers rarely restrict themselves to gathering one or two parameters. Each supplementary parameter represents an extra dimension. Even though a graphical representation can still be obtained for three parameters (Figure I.1(B)), a **multidimensional** graphical representation is no longer possible beyond this number. Graphical analysis then becomes laborious since it involves the analysis of the two-dimensional (2D) projection of all the combinations of parameters taken two by two in order to understand all the possible "relationships" between the parameters and the potential "similarities" between dates/stations in relation to these parameters. In this sense, numerical analysis will allow us to have an overview of the relationships between all the relevant variables and to determine how stations/dates are similar concerning the evolution of these parameters.

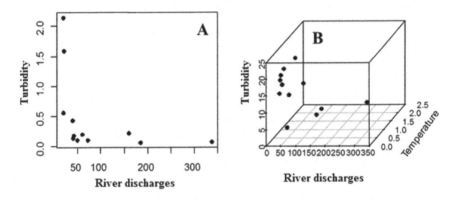

Figure I.1. *Graphic representation of data from the Charente estuary (fixed point with a salinity of 5). (A) 2D representation of turbidity in relation to river discharges. (B) 3D representation of turbidity in relation to river discharges and temperature*

The goal of environmental researchers will therefore consist of **summarizing the information** provided by their data sets:

1) by reducing the number of dimensions (i.e. correlated variables may provide redundant information);

2) by highlighting the similarities between parameters (e.g. positive/negative monotonic relationships and nonlinear relationships);

3) by identifying general trends/structures in the observations (is there a particular structure – gradient, groups – of dates/stations?);

4) to find out the causes behind these structures (i.e. which parameter(s) explain(s) the formation of groups or gradients?).

I.1.2. *Why this book?*

The objective of this book is to present the most used numerical analyses that allow us to reach these goals. Obviously, it will not be exhaustive, due to the large number of analyses in place right now and the constant development in this field.

There are two different schools of thought in relation to how these analyses are used:

– the Francophone school [LEG 98], which follows parametric approaches and increments the analyses in the R software;

– the Anglophone school [CLA 93], which favors non-parametric approaches and has developed the PRIMER software (see section I2.2 for the notions of parametric and non-parametric approaches).

Most of the time, researchers and teachers follow one of these two schools of thought. Scientific literature shows how the Anglophone school's adepts most often choose non-parametric analyses, even if their data would allow them to use more powerful parametric approaches in some cases, and how the Francophone school's adepts use parametric approaches without questioning themselves about their relevance in relation to their data sets.

The central concern of this book is thus to make both sets of analyses accessible, to make it possible to interpret them correctly and to provide some leads to help people find the most relevant analyses for the data sets used in relation to the scientific goals defined beforehand.

A graphical approach that considers parameters two by two is a mandatory preliminary step if we want to understand data sets. It is highly recommended before and in tandem with multivariate analyses in order to:

1) verify the quality of the data, i.e. to discover outliers and correct them if necessary (e.g. distinguishing between errors and extreme data);

2) ensure that the general trends observed are plausible, that is to say the numerical approach is correct and its interpretation is valid;

3) ensure that all the relevant pieces of information are highlighted in relation to the data set and achieve the predetermined scientific goals.

Using this kind of analysis allows us to have an overview of the results and enables us to summarize in one or two figures what would have required about 30 figures. The number of figures is actually limited for all internship reports and scientific publications. Moreover, these analyses can provide information that would not have been found by a simple analysis of 2D graphs.

I.1.3. Why the R programming language?

The R software is simultaneously a programming language and a workspace facilitating data treatment. It is used to manipulate data sets, draw graphs and carry out statistical analyses of these data.

Among its advantages, R (1) is a "free" program that can be downloaded and installed on all computers, (2) works on several operating systems such as Windows, Linux, Mac OS, etc. Consequently, scientists employ it internationally and share their statistical knowledge by developing new functions or packages and by interacting through forums. This program is thus rapidly and constantly evolving. Numerous statistical analyses are available in R, both simple and complex (i.e. descriptive and inferential statistics, parametric or non-parametric tests, linear or nonlinear models, numerical ecology, multivariate analyses, signal treatment and spatial analyses). Very few commercial pieces of software can freely offer such a choice of analyses. All the parametric and non-parametric approaches presented in this work are incremented in R. Finally, R also offers several very useful graphical functions.

This program is very effective in allowing us to carry out more sophisticated numerical analyses on the data (through a succession of functions or loops) and to reproduce them very quickly on other data sets by creating our own scripts.

I.2. General types of multivariate analysis

Multivariate analysis is an integral part of numerical ecology, i.e. the branch of quantitative ecology that deals with the numerical analysis of data sets [LEG 98]. It makes it possible to treat databases in which each sampling element (for example stations/dates) is defined by several variables (descriptors) in one go. Multivariate analysis simultaneously combines statistical and non-statistical approaches. **Two general types of multivariate analysis** are distinguished allowing one to move from the description to the explanation of a data set (i.e. exploratory and explanatory analysis), while **two kinds of approaches** can be used in relation to their power and robustness (i.e. parametric and non-parametric approaches).

I.2.1. Two general types of multivariate analysis

I.2.1.1. "Exploratory" analyses

The common goal of these analyses is to **describe the structure of a data set**. For example, they aim to determine how stations vary from one another based on the biological descriptors (e.g. phytoplankton species). There are **three general kinds** of analyses:

– Cluster analysis

This analysis aims to **classify objects** (e.g. date/stations) **or descriptors** (e.g. chemical–physical or biological variables) **into groups of similar units**. Thus, its goal is to identify discontinuities in a data set. The formation of these groups takes into account all the information contained in the data set. However, this kind of analysis will not allow us to obtain any pieces of information on the parameters that generate these groups on its own. If three groups of stations are highlighted by a classification, it means that the stations included in each of these groups are similar considering the set of parameters considered. However, we will not obtain any information about how and to what extent each parameter explains the difference between these groups.

Example: Hierarchical clustering.

– Multidimensional scaling

This analysis aims to **represent, in a reduced-dimension space** (for example a plane), **the similarities/differences between objects or descriptors**. It considers all the information included in a data set as for clustering analysis. It is based on the principle that environmental or biological data sets are structured in the shape of a gradient and that restricting objects into groups is difficult. Its ability to highlight groups or gradients makes it more powerful than clustering analysis. However, the possibility of using this analysis will depend on its ability to represent all the information included in a data set (e.g. 68 species) in a small number of dimensions (two or three). Finally, it exhibits the same shortcoming as clustering: it cannot provide information about how and to the extent to which each parameter generates the gradients/groups on its own.

Example: Non-metric multidimensional scaling (NMDS).

– Unconstrained ordination

This analysis aims to **identify and hierarchize the general trends in a data set in relation to the objects and the descriptors**. These general trends are represented by a restricted number of axes that summarize the information. Thus, all the information included in the initial data set will not be expressed by the few axes considered. In addition, the gradients or groups of elements (e.g. stations) will be related to the parameters that generate them. If the stations are distributed along the first two axes, and if the former is well correlated to the nitrogen or phosphorous concentrations and the latter to temperature, this will mean that the structure of the stations is predominantly explained by the availability of nutrients and, secondarily and independently, by temperature. Its ability to rank the information included in the data set **makes unconstrained ordination more powerful than cluster analysis or multidimensional scaling**. However, these analyses depend on certain applicability conditions, which make them **much less robust**, especially for small data sets (i.e. a small number of objects in comparison with the number of descriptors). Refer to section I2.2 for the terms "power" and "robustness".

Example: Principal component analysis.

I.2.1.2. *"Explanatory" analyses*

These analyses aim to **compare the structure of the data obtained through the aforementioned exploratory approaches with other data sets** in order to explain this structure. For example, highlighting a gradient of stations based on phytoplankton communities requires us to explain this gradient by means of environmental variables (for example chemical–physical parameters, hydrodynamics and land use). There are two ways of achieving this.

– A posteriori **(or indirect) analysis**

This analysis aims to **link the structure obtained (as it is) to potentially explanatory parameters,** either through the passive projection of supplementary variables or by correlating the latter parameters with the axes of an exploratory analysis or by relating them with groups defined *a priori* (e.g. by qualitative factors based on preliminary cluster analysis). This is a passive approach.

Example: Supplementary variables.

– A priori **(or direct) analysis**

The principle of this kind of analysis is to **constrain the structure before the analysis** by means of potentially explanatory parameters. This is an active approach. In order to become more effective, this analysis needs to limit the number of potentially explanatory variables to the relevant ones.

Example: Canonical correspondence analysis.

I.2.2. *Power and robustness: inferential statistics applied to multivariate analysis*

I.2.2.1. Inferential statistics applied to multivariate analysis

Some statistical tests are used in numerical analysis to verify in a significant way:

1) whether the structures highlighted (e.g. groups) are different;

2) whether certain descriptors can explain these structures;

3) whether these structures are linked to certain environmental variables;

4) the correspondence between structures highlighted by two different analyses.

Observation: This is the first time that the word "significant" has been used. Most numerical analyses can be interpreted only graphically (with some strict rules). In these cases, it is strongly advised to avoid describing the results obtained in terms of "significant" trend.

In simpler terms, inferential statistics is based on tests that allow us to say whether the results are "significant". In standard inferential statistics, these tests enable us to compare the estimators (i.e. means, variances and percentages) of a variable (e.g. temperature) between two or more populations based on random samples of them. The principle of parametric tests is to start from the hypothesis that the variable follows a known theoretical probability distribution (e.g. normal distribution) and, therefore, that the null hypothesis associated with the test follows this distribution. The problem of data tables is that they present several variables instead of just one. Thus, the statistical tests developed must integrate this multidimensional aspect and, consequently, the principle adopted will differ: the theoretical distribution upon which the test will be based will be calculated from random permutations of the initial data set, which generates randomness and, therefore, a distribution that corresponds to the null hypothesis (i.e. permutation tests).

Example: SIMPROF test.

I.2.2.2. *Parametric and non-parametric approaches applied to multivariate analysis*

Obtaining all the data elements of a population is utterly unrealistic, e.g. all the oysters of the Arcachon bay (due to temporal and staff-related constraints, analysis requiring the death of the analyzed individuals, etc.). Therefore, research involves **representative samples**, namely samples that reflect the complexity and the composition of the studied population. These representative samples can only be obtained by random sampling, whereby all the individuals of the population stand the same chance of being part of the sample. However, sampling fluctuation means that two samples belonging to the same population may give different results concerning a given parameter and that two samples taken from different populations may yield identical values concerning this parameter. Only statistics can allow us to extrapolate the values obtained from a sample to the whole population. In classical statistical tests, two working hypotheses are made: a null hypothesis and an alternative hypothesis. For example, if we are comparing two populations (e.g. considering temperatures), the null hypothesis (H0) assumes that sampling fluctuation accounts for the differences between the samples taken from the two different populations (i.e. no noticeable difference), while the alternative hypothesis (H1) admits that sampling fluctuation cannot explain everything (i.e. it is concluded that there are some differences). In most cases, environmental researchers aim to reject H0, namely to underscore a significant effect. Statistical tests allow us to decide in favor of one of these two hypotheses.

The **power** of a test is its ability to reject H0 when H0 is false, that is to say to highlight a difference between two populations (based on the samples) when this difference is real. **Robustness** is its sensitivity to deviations from the hypotheses made, namely its ability to provide reliable results when the applicability conditions are not respected. Thus, for each kind of analysis (e.g. the comparison between the mean of two independent samples), there will be at least two tests:

– **A parametric alternative**: More powerful but less robust, it will be more likely to reject H0 if H0 is false, i.e. to state that there is a difference between two samples if there is one since it is based on a known theoretical probability distribution. However, it will be more sensitive to the applicability conditions of the tests. If the latter are not respected, the results might be wrong.

– **A non-parametric alternative**: Less powerful but more robust, it will be less likely to reject H0 if H0 is false. However, it will present very few constraints in relation to the applicability conditions of the tests.

In relation to the two general types of multivariate analysis, some analyses can be said to be parametric, while others can be named non-parametric (Figure I.2). Parametric analyses will be more powerful in terms of the types of results provided, but they will be constrained by a certain number of applicability conditions. For example, this is true for unconstrained ordination, which allows us to hierarchize the information of a data set while also being constrained by the condition of data multinormality (i.e. each variable must follow a normal distribution). Other less powerful analyses will present no preliminary applicability condition. They are generally based on ranks (e.g. association coefficient ranks), rather than raw values. For example, this is the case for NMDS: only the similarities or differences between stations on a 2D plane can be analyzed without the ability to hierarchize the information. However, it presents no applicability condition, with the exception that the representation quality indicator (i.e. stress) has to be good in order for the interpretation to be correct.

I.2.2.3. *Consequences: different analyses for "small" and "large" data sets*

Thus, the approaches used (parametric vs. non-parametric) will depend on the size of the data set (Figure I.2). In general terms, parametric types of analysis (e.g. unconstrained ordination, canonical analysis) will be suitable for "large" data sets, whose number of elements (for example stations/dates) is large and especially greater than the number of descriptors (e.g. chemical–physical variables). On the other hand, non-parametric approaches, whose robustness is high, will be favored for "small" data sets (i.e. a small number of elements, which is smaller than the number of descriptors). Their power will be reduced (e.g. less information about ranking the significance of the descriptors in relation to the elements' structuration will be available), but the numerical analyses used will not be biased by the weakness of the data set (the drastic applicability conditions linked to the mathematical principles of the parametric approaches on which these analyses are based will not be necessary).

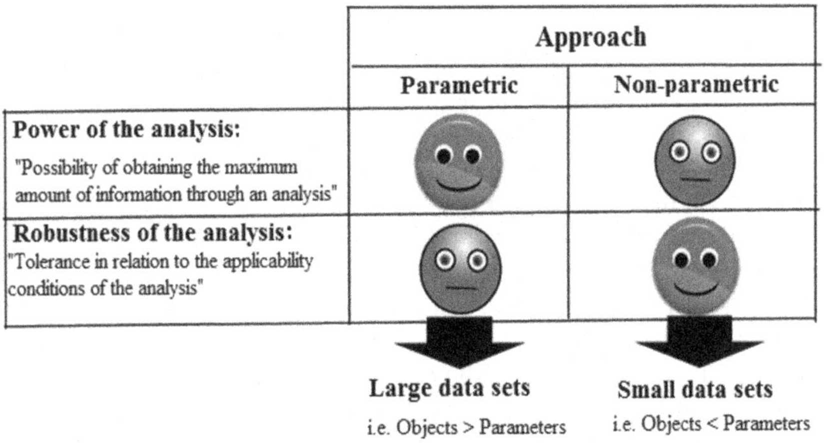

Approach		
	Parametric	Non-parametric
Power of the analysis: "Possibility of obtaining the maximum amount of information through an analysis"	☺	☹
Robustness of the analysis: "Tolerance in relation to the applicability conditions of the analysis"	☹	☺
	Large data sets i.e. Objects > Parameters	**Small data sets** i.e. Objects < Parameters

Figure I.2. *Parametric and non-parametric approaches applied to multivariate analyses. Conditions in terms of power and robustness and consequences in terms of applicability on small and large data sets (number of objects vs. number of parameters)*

In any case, less is more: non-parametric approaches can be applied to "large" data sets. However, if a parametric approach can be used to achieve the scientific goal, it is advisable to employ it in order to be able to obtain more information about the database processed (e.g. if applicable, an unconstrained ordination will provide information about ranking the significance of the descriptors in relation to the objects' structure, whereas using a non-metric multidimensional analysis will give only the objects' structuration, without providing any information about the role of the descriptors involved). Some analyses, however, allow us to adopt an intermediate approach that is more powerful than a non-parametric one and yet presents less constraints in terms of applicability conditions (e.g. permutational analysis of variance). Testing their suitability for a small data set is advised before moving on to a non-parametric approach.

I.3. Choosing analysis types in line with the research objective

In the best-case scenario, a sampling strategy or an experimental plan has been implemented to reach one or more specific goals. It is important to be already thinking about the kind of numerical analysis that will be carried out during the elaboration of the sampling process or of the experimental plan [SCH 84].

We should never lose sight of the objectives while data are being processed, lest we start carrying out analyses for the sake of carrying out analyses, so that we end up not answering the questions asked at the beginning of the scientific study.

Several analyses will allow us to answer the same question. What matters is to choose the most relevant analysis (or analyses), which provides as much information as possible insofar as the applicability conditions are respected. If several analyses are carried out, they must be complementary, since carrying out redundant analyses is useless.

We should take a critical look at the analyses put forward in the literature. These analyses do not have to be necessarily open to criticism, but just because a researcher uses a specific analysis to achieve the same goal, it does not mean that it is advisable to carry out the same analysis. His or her sampling strategy may be different (e.g. number of stations, types and numbers of parameters, etc.) and, consequently, the analyses that should be carried out may vary, for example, due to the applicability conditions.

Similarly, when a database is taken from another study, we must make sure that the sampling strategy or the experimental plan adopted beforehand is suitable for the goals we have set.

I.3.1. *Is it possible to manipulate the results of the analyses with such tools?*

Numerical analysis remains first and foremost a tool that should be used rigorously in full knowledge of the data and, especially, the way in which it has been obtained. A researcher would not think (1) of using an oxygen probe that has not been previously calibrated or (2) of identifying living organisms without employing specific determination criteria listed in specialized identification keys. The false but widespread adage that "statistics can make data say anything or whatever we want" clearly shows that these analyses have been misread.

I.3.2. *"To make data say anything": Which mistakes can lead us to interpret data incorrectly?*

Most analyses are based on strong hypotheses that affect the properties of the mathematical tools developed in relation to them. For example, unconstrained ordination starts from the principle that data follows the normal distribution for all the descriptors considered and that the interpretation that will be made is only and rigorously valid if this condition is respected. **Respecting the applicability conditions,** therefore, guarantees that the interpretation will be valid for all these analyses.

Interpreting these analyses correctly by being aware of their limits is equally important: each of them presents its own interpretation rules, which derive especially from the properties of the mathematical tools used and from the hypotheses made beforehand. Thus, we must be aware of what we can and cannot conclude with the analysis chosen. For example, we will not be able to interpret descriptors that are not adequately represented on the axes of an unconstrained ordination under any circumstances, since we run the risk of drawing the wrong conclusions.

Another especially significant constraint of these analyses derives from the fact that **they are only mathematical tools**. The results they provide must be considered **in a critical way by a researcher who remains an expert in his/her field**. Empirical or bibliographic knowledge of the field is of great help to any interpretation. Highlighting a correlation between two or more variables does not guarantee a cause–effect relationship since relationships between variables are especially complex.

– **The correlation between two variables may be coincidental.** Let us imagine that a strong correlation has been highlighted between the land use percentages of a drainage basin employed as agricultural land and forestland, and the quantity of nitrates in the downstream fluvial waters. If it is evident that high concentrations of nitrates can be explained in all likelihood by the significant presence of agricultural land, rather than large areas of forestland, in the drainage basin, only empirical knowledge of the field can lead us to the conclusion that the relationship between agricultural land and forestland is a mere coincidence in the geographical area considered.

– **Two variables can be independently explained by the same variable**. The correlation between two variables can only result from the effect of a third one. Let us consider the example of two species, a benthic and a planktonic one for which abundances are strongly linked to temperature: high temperatures favor the benthic species, whereas low temperatures stimulate the planktonic taxa. From this correlation with temperatures, we may deduce a pronounced anticorrelation between the two species, even if there is actually no real negative relationship between them that might be explained by predation, competition, etc. Only in-depth knowledge about these two species allows researchers to identify whether this negative relationship is real or caused by the independent link between each of them and temperature.

– **A relationship between two variables may be positive or negative according to the scale considered.** Temperature may be a parameter that affects the physiology of a species in a positive way on a small scale (seasonal scale) and negatively on a long-term scale (decadal scale). For example, temperature is a positive factor acting on the physiology of poikilothermic species and, therefore, it controls their seasonal cycle. However, a long-term increase in temperature may amplify the thermic difference between summer maximum and winter minimum temperatures, thus becoming harmful to these species.

– **Several variables may account for the fluctuations of another variable (e.g. a species) and these independent variables may also be intercorrelated**. An increase in temperature may cause a stratification of the water column and, consequently, a more rapid depletion of the dissolved nutrients with a negative effect on diatoms (planktonic microalgae). Even if certain species of diatoms may be favored by increasing temperatures, the temperature threshold, which may or may not lead to a stratification, will favor or hinder the development of this taxon.

I.3.3. *"To make data say whatever we want": Does this imply that we choose the analysis that suits us out of several of them that provide different results?*

Of course not! A rigorous approach will allow for several analyses, but the results will be coherent, redundant or complementary. Choosing the analysis to consider will deal with the quantity and quality of the information provided in relation to the predetermined scientific goal, while avoiding presenting totally redundant analyses.

A rigorous approach requires **some choices to be made** while preparing the data set and throughout the numerical analysis.

– A data set may, for example, present missing data, which might be treated according to its type: an absence of data that represents the absence of a species may be replaced by 0, while an absence that represents a machine failure that prevents us from taking the measurement must be identified as missing data (written as NA in R language).

– Data may require a transformation that is chosen in relation to its type and the requirements related to the analysis (e.g. standardizing units of measurement, "normalizing" data and reducing the range of variations).

– The objectives and the way data has been obtained will govern the choice of the association measure, which allows us to measure the degrees of similarity between objects or descriptors. For example, the coabsence of a species in two stations can be considered as a criterion of similarity between stations if this coabsence represents a specific phenomenon that we are trying to study (e.g. pollution), whereas it can be ignored if we do not want to give it too much weight (e.g. when the sampling method is biased by rare species).

All of these choices will affect the relationships or similarities observed between objects (i.e. stations/dates) or descriptors. It is clear that, in order to reach a certain goal, these choices will be the same regardless of which analysis is carried out. Consequently, results will be coherent in all the approaches used, even if some analyses may provide some variants according to their specific characteristics and mathematical properties.

I.4. The database used as an example in this book

The data presented here will be employed throughout the text to show a practical application of the analyses used. For each objective described below, we will mention the chapters in which the relevant analyses will be carried out.

Four drained marshes in the Charente-Maritime department (South-West France) have been sampled. These marshes have been created by man over the centuries for the development of agricultural activities (i.e. agriculture, livestock farming, oyster-farming, fish farming, salt production, etc.) and they have also been subjected to a progressive urbanization (sewage plants, houses). They represent a hydrographic artificial network of channels and ditches that range from 10 to several hundred kilometers in length. Human intervention, through them, aims to control lock gates on the coast to prevent saltwater from penetrating into the network with every incoming tide and to avoid the flooding of nearby land during winter floods. The water system is made of three types of channel: (1) large main channels, deeper than 1 m, in which water flowing in from other channels is drained before reaching the coastal waters; (2) ditches or tertiary channels – they are the narrowest type of channel and they are no deeper than 50 cm – which directly supply the land nearby where anthropogenic activities develop; and (3) intermediate or secondary channels, which ensure that water is drained from ditches all the way up to the main channels.

The data are available as an excel file (Data_Marsh_book.xls) on the Researchgate profile of the author (https://www.researchgate.net/profile/V_David). Each of the databases described below is presented as a folder.

The four marshes (A–D) have been sampled during the summer (in July): two drained marshes re-fed by the river nearby (the Charente) to prevent summer droughts, when farmland needs particularly high levels of water (R), and two unfed drained marshes whose water level can quickly decrease, in summer, when some ditches dry up completely. For each of these kinds of marsh, sampling was processed inland (I) and on the coast near sea lock gates (E). For each of the four marshes, four to six stations have been sampled (written as "a" to "e") for different kinds of channels and different land uses on neighboring areas of influence. For each of the 19 stations, different parameters have been sampled and marked (see Table I.1). Each folder of the excel file must be imported in R using the read_excel() function of the readxl library:

– **Environmental parameters (folder "fext")**: Station (variable "Station", from A to D); type of marsh (variable "Type", D or R); "position" (I for internal and E for external); land use on the neighboring area of influence (variable "use", with Grassland, Farmland or Urban); types of channel (variable "Channel", with "prim" for primary, "sec" for secondary and "ter" for tertiary); presence of macrophytes (variable "MP", with "yes" if macrophytes are present) and surface of the drainage basin (variable "DB" in km^2). The codes for the stations appear as a name arranged in a row with the type of marsh (from A to D) and the stations sampled in these marshes (from a to f).

– **Chemical–physical parameters (folder "CP")**: Depth ("Depth"); solar radiations at the water surface ("Lumin"); optical depth ("Opt_depth"); water temperature ("Temp"); concentration of nitrites ("NO2"), nitrates ("NO3") and phosphates ("PO4"); the ratio (NO3 + NO2)/PO4 ("N.P"); and turbidity ("Turb").

– **Global biological parameters (folder "bio")**: Chlorophyll biomass as an index of the phytoplankton biomass ("Chloa"), primary production in the water column ("P_phyto"), phytoplankton productivity–production/biomass (P.B) ratio.

– Phytoplankton abundance (folder "phyto"): The abundances of 68 genera have been listed, but only the abundances of seven genera are provided in Table I.1. The whole 68 genera and their respective codes are listed in Table I.2.

library(readxl)

Verify that the compatibility with your excel version (put "," as decimal
   separation for a French version in your Data_Marsh_book file for
   example...) and change the directory

fext <- read_excel("~/R/Data_Marsh_book.xlsx", col_types = c("text", "text",
   "text", "text", "text", "text", "text", "numeric"), sheet = "fext") # *Import de*
   first sheet "fext"

rownames(fext)<-fext[,1]; fext<-fext[,-1];summary(fext) #*To use the Code column*
   as rownames

fext$Station<-as.factor(fext$Station);fext$Type<-as.factor(fext$Type);

fext$Position<-as.factor(fext$Position); fext$Land_use<-as.factor(fext[,4]);

fext$Channel<-as.factor(fext$Channel);fext$MP<-as.factor(fext$MP) #*To change*
   the qualitative variables as factors

fext[,-4]->fext # *To eliminate de Land use column that had been replaced by*
   Land_use

as.data.frame(fext)->fext # *To give the good format to the database*

summary(fext) # *To verify the successful of the import and the kind of variables*

CP <- read_excel("~/R/Data_Marsh_book.xlsx, col_types = c("text", "numeric",
   "numeric", "numeric", "numeric", "numeric", "numeric", "numeric",
   "numeric","numeric"), sheet = "CP");

rownames(CP)<-CP[,1]; CP<-CP[,-1];summary(CP)

bio <- read_excel("~/R/Data_Marsh_book.xlsx", col_types = c("text", "numeric",
   "numeric", "numeric"), sheet = "bio"); rownames(bio)<-bio[,1]; bio<-bio
   [,-1];summary(bio)

phyto <- read_excel("~/R/Data_Marsh_book.xlsx",sheet="phyto");rownames(phyto)
   <-phyto[,1]; phyto<-phyto[,-1];summary(phyto)

Code	Station	Type	Position	ENVIRONMENT Land use	Channel	MP	DB	Depth	Lumin	Opt_Depth	CHEMICAL-PHYSICS Temp	NO2	NO3	PO4	N.P	Turb	BIOLOGY Chlaa	P_phyto	P.B	PHYTOPLANKTON Cycl	Melo	Licm	Frag	Nitz	Cosm	Eugl	
A.a	A	D	I	Prairies	sec	Oui	4716	0.5	856.5	0.35	23.25	0.01	0	3	0.003	11.6	14.40	24.17	0.026	305	56	0	0	28	0	0	--
A.b	A	D	I	Cultures	prim	Non	4716	0.7	1886.25	0.05	30	0.018	0	0.1	0.18	23.5	21.59	2640.81	1.912	0	411	0	0	17625	137	1646	--
A.c	A	D	I	Cultures	ter	Oui	4716	0.3	1940.5	0.05	25.5	0.028	0	2	0.014	38.3	87.45	13703.69	2.448	805	0	0	0	889	611	666	--
A.d	A	D	I	Cultures	prim	Non	4716	0.8	1704	0.15	29.8	0.021	0	1	0.021	34.9	43.88	3815.07	1.360	154	0	0	0	463	695	116	--
A.e	A	D	I	Cultures	prim	Non	4716	0.25	346.4	0.02	26	0.091	0	1	0.091	58.5	95.25	7395.07	1.213	0	0	0	0	4860	194	889	--
A.f	A	D	I	Cultures	ter	Non	4716	0.08	817.5	0	26.4	0.073	0	2	0.0365	60	189.48	18768.42	1.548	41098	0	0	0	9025	1944	2778	--
B.a	B	D	E	Prairies	prim	Oui	6971	0.48	1761	0.15	25	0.049	0	1	0.049	60	128.04	12740.47	1.555	15984	0	0	0	224	671	559	--
B.b	B	D	E	Prairies	sec	Oui	6971	0.3	1833.5	0.27	24.5	0.015	0	1	0.15	48.6	65.99	8524.27	2.018	139	0	0	0	333	0	1186	--
B.c	B	D	E	Urbain	prim	Oui	6971	0.75	132.5	0.3	24	0.05	1	3	0.35	60	330.08	697.64	0.033	11108	0	0	0	12774	6109	555	--
B.d	B	D	E	Prairies	sec	Non	6971	0.33	1535	0.08	24	0.011	0	2	0.006	14.4	12.50	1946.69	2.433	139	139	0	0	208	0	417	--
C.a	C	R	E	Prairies	ter	Oui	1325	0.3	1306	0.15	22	0.003	0	1	0.03	10	23.97	4400.87	2.868	4	30	69	0	3	0	0	--
C.b	C	R	E	Urbain	sec	Non	1325	1	1899	0.07	22	0.013	0	1	0.13	40.2	45.90	4287.45	1.460	28	0	0	250	14	167	458	--
C.c	C	R	E	Urbain	ter	Non	1325	0.77	1426.5	0.47	22	0	0	2	0	13	20.73	1638.16	1.235	4041	0	0	0	458	14	625	--
C.d	C	R	E	Urbain	prim	Non	1325	1.5	744.5	0.45	23	0.013	0	3	0.004	53.4	196.90	1236.26	0.098	0	0	0	0	19300	0	42488	--
D.a	D	R	I	Urbain	prim	Non	982	1.33	1781.5	0.04	26	0.091	2.7	1	27.91	60	36.02	859.92	0.373	2	91	0	0	132	1	5	--
D.b	D	R	I	Prairies	ter	Non	982	0.53	1854	0.1	25.5	0.07	2.5	1	25.7	60	345.56	27496.77	1.243	112	72	0	0	52	0	2462	--
D.c	D	R	I	Cultures	prim	Oui	982	0.93	2128.5	0.2	24	0.039	2.9	1	29.39	60	55.84	4500.68	1.259	695	0	0	0	144	0	1228	--
D.d	D	R	I	Prairies	ter	Oui	982	0.85	1310	0.33	23	0.036	2.9	1	2.936	55.2	60.70	3611.1	0.930	0	0	0	0	689150	0	7980	--
D.e	D	R	I	Prairies	sec	Oui	982	0.83	1887.5	0.35	23.5	0.031	2.9	1	29.31	44.4	19.12	1989.24	1.626	28	0	0	0	6	0	1028	--

Table I.1. *Environmental, chemical–physical and biological parameters of the samples taken at 19 stations in the marshes in the Charente–Maritime department. Only the abundances of seven taxa of microphytoplankton are given (out of the 68 total)*

Taxon	Code	Taxon	Code
Cyclotella	Cycl	Staurodesmus	Staud
Melosira	Melo	Ac nastrum	Ac
Licmophora	Licm	Ankistrodesmus	Anki
Fragillaria	Frag	Crucigenia	Cruc
Achnan nium	Achnm	Haematococcus	Haem
Meridion	Merid	Monoraphidium	Mono
Synedra	Syne	Oocys s	Oocy
Achnanthes	Achn	Pandorina	Pand
Rhopalodia	Rhop	Pediastrum	Pedi
Caloneis	Calo	Scenedesmus	Scen
Gomphonema	Gomp	Desmodesmus	Desm
Gyrosigma	Gyro	Tetrastrum	Tetrs
Haslea	Hasl	Tetraedon	Tetrd
Navicula	Navi	Chroomonas	Chroas
Pleurosigma	Pleu	Cryptomonas	Cryp
Amphora	Amph	Chrysochromulina	Chry
Cymbella	Cymb	Chlamydomonas	Chla
Cymatopleura	Cyma	Tetraselmis	Tetrl
Epithemia	Epit	Euglenes	Eugl
Bacillaria	Baci	Lepocinclis	Lepo
Cylindrotheca	Cyli	Peranema	Pera
Nitzschia	Nitz	Phacus	Phac
Pseudonitzschia	Pseu	Strombomonas	Stro
Gymnodinium	Gymn	Trachelomonas	Trac
Peridinien	Peri	Anabaena	Anab
Gloecys s	Gloe	Gomphospaeria	Gompp
Cystodinium	Cyst	Chroococus	Chro
Gonyostomum	Gony	Merismopedia	Meris
Synura	Synu	Mycrocys s	Mycr
Closterium	Clos	Nostoc	Nost
Dinobryon	Dino	Plankto	Plan
Netrium	Netr	Oscillatoria	Osci
Cosmarium	Cosm	Spirulina	Spir
Staurastrum	Staut	Diversecyano	Cyan

Table I.2. *Phytoplankton taxa listed throughout the study and abbreviations used for the following analyses*

I.5. The structure of this book

This book is organized into five chapters.

Chapter 1 (observing and preparing a data set): The goal of the first part of this chapter is to find out more about the data set, i.e. to establish its boundaries and make the necessary choices before any kind of treatment, in keeping with the goals of the study and the sampling strategy (sections 1.1 and 1.2). It also describes the methods used to simplify data sets in order to remove redundant or uninformative variables (see sections 1.3 and 1.4).

Chapter 2 (preliminary processing of the data set): This chapter aims to show how to calculate diversity indices (see section 2.1), how to transform the data set (see section 2.2) and how to choose the most suitable association measure for the data set (see section 2.3).

Chapter 3 (clustering objects/variables): This chapter describes the most used clustering methods, explains how to apply them in relation to the data set and the objectives (see section 3.1) and shows how to define the groups obtained (see section 3.2).

Chapter 4 (Gradient of objects/variables): This part describes the application of an unconstrained ordination as a parametric approach (see section 4.1) and the application of non-metric multidimensional scaling as a non-parametric approach (see section 4.2).

Chapter 5 (understanding a structure): This part aims to present methods that directly compare structures with no hypotheses in relation to independent and dependent variables (see section 5.1), and to find quantitative and qualitative factors (see section 5.2) that account for the structures obtained in the previous chapters.

Figure I.3 illustrates how the analyses presented in this book are used in relation to the properties of the data set employed.

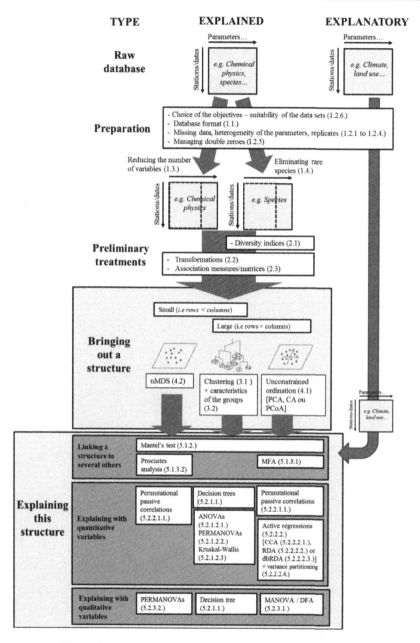

Figure I.3. *Diagram that allows us to choose the analyses presented in this book in relation to the data set and the goals*

1

Observing and Preparing a Data Set

This chapter aims to find out more about a data set, namely to establish its boundaries and to make the necessary choices before any kind of treatment, in keeping with the objectives of the study and the sampling strategy (sections 1.1 and 1.2). It also illustrates how to simplify data sets so as to remove redundant or uninformative variables for environmental factors (see section 1.3) and biological data (see section 1.4).

1.1. Creating a database in relation to our predetermined goals

During a sampling process, researchers take their samples in several sampling points (i.e. spatial sampling), on different dates (i.e. temporal sampling), or in both ways (i.e. spatiotemporal sampling). In the database, these stations or dates will represent the objects. For each object, researchers will sample different kinds of parameters in relation to their discipline and objectives: we may deal with ecological samples (e.g. fauna or flora sampling), environmental samples (temperature, wave height, current, etc.), or samples concerning the phylogenetics of a given species – i.e. morphological (for example wing length, size, weight, etc.) or molecular (e.g. DNA and proteins) features. If several replicates are available in each station or on each date, each one of them will represent a different object in the raw database, called a **sampling element or sampling unit**. Each element will be arranged by convention in a row in the raw database and is identified by a code that varies for each object, even if it has been taken as a sample at the same station/on the same date (e.g. the sampling of 20 oysters taken in one station: each oyster is listed in a row and, consequently, there will be 20 rows identified for the same station).

Descriptors or variables will be the characteristics measured or observed for each element of the sample, according to the kind of sampling. Variables may be:

– *quantitative* (including measurable elements):

- *continuous*: all values are allowed (e.g. temperatures and concentration of nitrates, Table I.1);

- *discrete*: exclusively with integer values (e.g. number of phytoplankton cells per liter, Table I.1).

– *Qualitative* (including values that express a quality):

- *nominal*: without logical gradation (e.g. code "Marsh", Table I.1);

- *ordinal*: with logical gradation (e.g. "Type of channel", Table I.1);

- *semiquantitative*: grouped by type of values (e.g. classes of abundances);

- *binary*: expressed as 0 or 1 (e.g. the presence or absence of macrophytes, Table I.1).

On the one hand, one of the objectives of a specific work may involve understanding or predicting the variations of one or several variables, which will be called **dependent or explained variables**. On the other hand, **independent or explanatory variables** are those variables that have an influence or effect that we try to underline. For example, phytoplankton production (dependent variable) could be explained in relation to the concentration of nitrates (independent variable).

Researchers may want to underscore the effect of one or several variables on one or several other variables through an experimental approach. In this context, **control variables** are parameters whose values are fixed by researchers, who try to determine their effect. On the other hand, variables that can take any possible value – especially the dependent variables – will be called **random variables**. For example, researchers may want to assess the impact of the concentration of a fertilizer (control variable) on the productivity of a cornfield (random variable).

During field sampling, variables take random values (e.g. phytoplankton production, concentrations of nutrients, etc.) unless some of them are fixed on purpose by a researcher in his sampling strategy. For example, salinity may be either a control variable when samples are taken at fixed salinities, for example at 0, 5, 15 and 25 in an estuarine system (Lagrangian sampling) or a random variable for a Eulerian sampling realized at fixed stations.

1.2. Observing the database scrupulously

This step is essential to direct the multivariate analyses that will be covered later on.

1.2.1. *Missing data*

The acquisition of a database is never perfect. Most of the time, raw data sets involve empty cells, which may be linked to a lack of data acquired in the field or bad-quality values due to an incorrect laboratory process. On the contrary, an empty cell in a flora or fauna database is often linked to the fact that a species has not been inventoried in the sample. The latter missing data correspond to an absence of the species and, consequently, to a value of zero, whereas the former missing values are actually missing and cannot be assessed. It will be expressed as NA (in R software). In any case, before downloading the database in R, filling all the empty cells is necessary and the previous issues (absence (0) vs. unassessed (NA)) must be tackled.

Multivariate analyses will not take into consideration the rows where data are missing. Thus, it will be wiser to remove from the database those variables containing too many NA values (even if this action may sometimes seem frustrating). Otherwise, the result of the analysis would be based on such a small number of rows that the results would not be representative of the sampling.

Sometimes it is possible, even if risky, to substitute the NA of some variables with values by replacing them (1) with the average of all the values of the variable, (2) with the mean between the previous and the following data in a process of seasonal monitoring and (3) by using methods that can regularize data with more precision by considering a larger spectrum of the data for a long-term temporal sampling process (e.g. splines).

It is strongly advised to carry out this task diligently before downloading the file in R, so that we no longer have to deal with the new raw data table (apart from the regularization processes).

1.2.2. The heterogeneity of the variables

The data set considered as an example includes most of the **different kinds of variables:** quantitative variables (continuous and discrete) and qualitative variables (nominal, ordinal and binary). The analyses employed for these several kinds of variables might differ and transformations may be necessary.

On the one hand, quantitative variables **are not necessarily** expressed in **the same units,** which implies that there are differences in terms of their range of variation. In the chemical--physical (CP) database (Table I.1), some variables, such as optical depth, present very limited ranges of variation (from 0 to 0.47 m) and values, whereas others, like solar radiation, present high values and wide ranges of variation (from 133 to 2,129). These differences are linked to the differences in terms of units. A different unit would have entailed changes in the values for a same variable: for example, if optical depth had been expressed in centimeter, the ranges of variation would have changed between 0 and 47. Nonetheless, the wide ranges of variation and high values of a parameter will have more weight in multivariate analyses due to the effect of the association measures on the calculations formulae. Thus, **a transformation will be necessary to eliminate the problem of heterogeneous units**. In this case, the most suitable transformation will be standardization, which allows us to reduce all the variables to the same unit (see section 2.2.1).

On the other hand, a database such as that of phytoplankton abundance presents *homogeneous units* (cells L^{-1}). However, the variation ranges of abundances from one species to another may be very different (Table I.1): *Nitzchia* varies between 0 and 689,150 cells L^{-1}, whereas *Licmophora* fluctuates between 0 and 69 cells L^{-1}. Besides, while the former taxon is inventoried at all stations, the latter has been listed only once and with low abundance. First of all, the step that involves the removal of rare species might allow us to reduce the data set in order to eliminate taxa like *Licmophora*, which are not very present or abundant among the 68 taxa listed (see section 1.4). However, there will still be marked differences in terms of variation range among taxa with a significant presence. For example, the maximum abundances of taxa with a high presence, *Nitzchia* and Euglena (Table I.1), are separated by a factor of 10. A transformation is thus required, as is the case for the CP parameters to **avoid giving too much weight to taxa with particularly high abundances**. However, this type of transformation will be different because (1) units are homogeneous, (2) 0s traducing an absence must be conserved as null values and (3) the transformation aims to reduce the range of variation; several transformations can be used in relation to the strength of the reduction that might be applied to the data set (e.g. square root and logarithm transformation; see section 2.2.2).

Moreover, some analyses (i.e. unconstrained analysis) require multinormal data and some transformations, such as standardization, might allow us to respect these applicability conditions. However, we cannot expect any miracles.

Besides, a homogenization of the data might sometimes be required through data degradation, because the quantification mode has not been obtained with the same methods through the sampling process. For example, the abundances inventoried in the phytoplankton database could be reduced to a simple presence/absence base of the taxa (section 2.2.2.2).

Finally, qualitative data cannot be used as they are in multivariate analysis. A transformation into binary variables will be required (see section 2.2.3).

However, it is important to point out that all these database transformations or degradations involve a **more or less significant distortion of the raw data** and will not provide the same results yielded by an analysis carried out on the raw database. Thus, these choices must be made beforehand in relation to the objectives and the constraints required by the available databases. The several types of most common transformations, as well as the distortions involved, will be tackled in section 2.2.

1.2.3. Dealing with replicates

Occasionally, it is useful to **summarize replicates** (e.g. within a station) in order to obtain only one value (per station). Let us reconsider the example of 20 oysters sampled for each station. It may be useful for certain analyses to reduce the 20 rows to one value per station, i.e. the mean or the median, before carrying on the analyses. This is not the case for our database, since the replicates represent, for each marsh, different stations in terms of environmental factors, and since between-station variability within a single marsh is interesting in itself.

To illustrate how to carry out this operation, the "chemical--physical" data table will be reduced to one value per marsh. The functions *group.by()* and *summarize()*, belonging to the package [dplyr], allow us to retrieve for each marsh the medians for every variable of the CP parameters.

```
library(dplyr)

by_marsh<-group_by (CP, fext$Type)

permarsh <- summarize (by_marsh, median(Depth), median(Lumin),
    median(Opt_Depth), median(Temp), median(NO2), median(NO3),
    median(PO4), median(N.P), median(Turb)) # Choosing the relevant
    variables and function

colnames(permarsh)<-colnames(cbind(fext$Type, CP)) # Changing the name of the
    variables in order to simplify them
View(permarsh)
```

The file "permarsh" stores only one object per marsh, which corresponds to the median values of all the replicates per marsh.

1.2.4. *The number of objects and descriptors*

For the most powerful kinds of multivariate analyses (unconstrained ordination or canonical analysis), the number of variables (descriptors) should theoretically be 10 times smaller than the number of objects. However, it is usually acceptable for the number of variables to be merely smaller than the number of objects (see section 4.1.1) provided that the number of objects is large enough (not fewer than 10 objects). The dimension of a data table is verified with the function *dim()*.

```
dim(CP)
```

[1] 19 9 # *19 objects / 9 variables*

Out of the four databases, the "chemical--physical" (CP), "environmental" (fext), and "biological without microphytoplankton" (bio) bases present fewer variables than stations sampled (9, 7 and 3 parameters, respectively, vs. 19 stations). In contrast, the "microphytoplankton" (phyto) base presents a number of variables that is more than three times greater than the number of stations (68 taxa vs. 19 stations). Consequently, it will not be possible to carry out an unconstrained ordination on it as it is, unlike the others. For this base, it would be more advisable to adopt a non-parametric approach (e.g. a non-metric multidimensional scaling, see section 4.2) or to reduce the number of taxa by eliminating rare species (see section 1.4) before carrying out unconstrained ordination (see section 4.1).

1.2.5. *Managing double zeroes*

When we try to assess whether two stations or dates are similar in terms of CP or biological descriptors, it may happen that values are equal to 0 for two elements (stations/dates) for the same variable. If we consider CP parameters, 0 has an actual value on the scale of the descriptor. A temperature of 0 °C, for example, is as meaningful as a temperature of 20 °C or −12 °C. The double 0, for both objects, adds a similarity between the two objects in relation to this parameter. Thus, it must be taken into consideration to assess the similarity between these objects.

If we are dealing with a matrix such as a flora or fauna inventory (for example phytoplankton), 0 implies the absence of the species considered, whereas the copresence of a species can indicate similar conditions in terms of ecological niches, and presence/absence may oppose two niches, a double absence (double 0) may be the result of different things (rare and thus uncollected species, sampling outside ecological niches for the two species, etc.). This double zero, linked to the joint absence of the two species at the two stations, is usually discarded when we assess the similarity between two stations, especially if our aim is to characterize biological communities. However, this coabsence of species may be essential for assessing certain goals, for example when we are looking for pollution indices.

The issue of whether we should consider or reject "double zeroes" during the analysis is thus of paramount importance, especially concerning the choice of association measures (see section 2.3).

1.2.6. *Goals/data sets compatibility*

The databases acquired in our example have the elements (stations) arranged in rows and the descriptors arranged in columns (CP parameters, phytoplankton taxa, etc.).

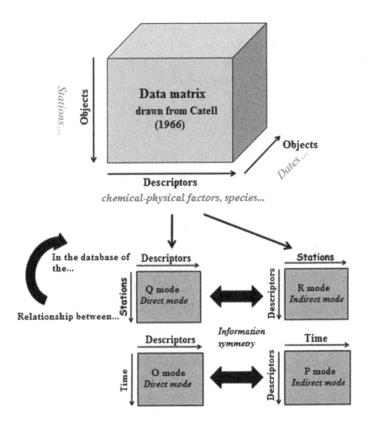

Figure 1.1. *Database and Q, O, R, P analyses modes, or direct versus indirect*

The table that will be used to apply a multivariate analysis will be oriented in relation to the goals we have set beforehand. For example, a data set with stations arranged into rows and descriptors arranged into columns will allow us to determine the relationship (i.e. similarities/differences) between stations on the basis of the descriptors. If the matrix is transposed, i.e. if the descriptors are arranged into rows and the stations into columns, it will highlight the relationships between descriptors. In the literature, we refer to **Q mode and R mode**, respectively [LEG 98; Figure 1.1). The two issues are different and therefore the type of results will also be different . In the former analysis, groups or gradients of stations will be underscored, whereas in the latter approach groups or gradients of descriptors will be highlighted. When the objects consist of dates, we will refer to **O mode and**

P mode, respectively (Figure 1.1). The Anglophone school refers to **direct mode** when we are generally looking for similarities/differences between objects on the basis of descriptors, and to **indirect mode** when we are looking for relationships between descriptors (Figure 1.1).

The association measures will vary in relation to the mode employed (see section 2.3). Determining our goal before we start carrying out analyses is important. Databases can be transposed with the function *t()* in R.

PCtranspose<-t(PC)

1.3. Reducing the number of environmental variables

1.3.1. Why is it relevant?

After being out in the field, researchers can often rely on a great number of environmental variables (or descriptors). Taking all these variables into consideration involves the risk of making the calculations complex and muddling the essential information conveyed by the database. Thus, it is important to detect the variables that play a minor role in the representation of the structure of the objects (stations/dates) and, consequently, to be able to choose whether or not to consider them or to discard them. Besides, some descriptors may be redundant if they are strongly correlated and they may give more weight to the information they provide. Finally, the smaller the number of descriptors in relation to the number of objects in the base, the more powerful the exploitable multivariate analyses. Here, we will use two methods. They may supplement each other in terms of information conveyed and interpretation of the final results: Escoufier's method (see section 1.3.2) and Spearman's correlation matrix (section 1.3.3).

1.3.2. Escoufier's method: reducing variables while keeping the maximum amount of information

Escoufier's method [ESC 70] allows us to extract a number of variables that is smaller than the initial one, so that the information kept and the structure of the objects obtained with these variables after a multivariate analysis resembles as closely as possible the same analysis carried out with all the variables. We obtain a database where variables are ranked in descending order in relation to the dependent variation of the initial table. The RV between-series correlation coefficient, which varies between 0 and 1, measures cross relationships between tables: the initial database and the different simplified bases. The method proceeds step by step by determining first the variable with the highest RV, then it keeps this variable and adds the others, one by one, to determine which combination of the two improves the RV, and so on.

Escoufier's method can be used with the function *escouf()* available in the package [*pastecs*] by specifying the database (descriptors arranged in columns and objects in rows) and the argument VERBOSE = TRUE to display all the details in the results (i.e. the order in which the descriptors are integrated when the process calculates the RV: the highest one and then, one by one, those that improve it). Here, it is applied to the matrix of the CP variables, which contains nine variables that vary in terms of variation range and units for the 19 stations.

library(pastecs)

CP.esc <- escouf(CP, verbose=TRUE)

summary(CP.esc) # *Summary of the analysis with the RV's for the descriptors in descending order of significance*

CP.esc$vr # *Correspondences between the number of a variable of the initial base in its output order and its name*

CP.esc$RV # *Correspondences between variables and related RV's in their output*

plot (CP.esc, main="Escoufier's vectors") # *Graphic representation of the RV (in black) and RV' (derivative of RV)*

This graphical representation of the Escoufier coefficients allows us to choose the number of variables considered. The best number corresponds to the break in the slope of the RV' associated with the threshold beyond which the addition of variables no longer explains anything significant.

CP.esc$level <- identify(CP.esc) #*Click then on the base of the break in slope of the RV'*

Level: 0.9312664 # *The 5 variables considered out of the total 9 account for 93% of the variation explained by the raw table.*

CP2 <- extract(CP.esc) #*Extraction of the variables selected from the database*

dim(CP2) #*Dimensions of the new database (there are only 5 variables)*

names(CP2) #*Names of the variables considered*

This method has allowed us to keep five variables (out of the total nine), which include 93% of the information provided by the initial CP database (Figure 1.2). These are mainly nutrition-related variables for phytoplankton (N/P ratio, nitrites and phosphates), the depth of the water column and temperature. The database containing the variables selected is stored in the object "CP2". This method enables us to compress the information of a table and to determine the variables that give us the largest amount of information about the structure of the objects (stations).

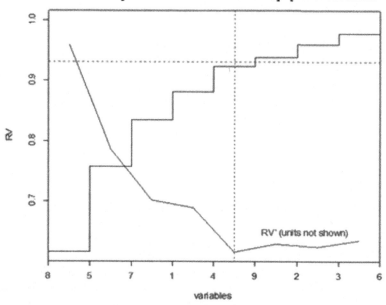

Figure 1.2. *Escoufier's method. Evolution of the RV coefficients and RV scree plot allowing us to choose the variable number to be kept in order to explain the maximum amount of information of the initial database without multiplying the parameters for further multivariate analyses*

1.3.3. *Spearman's correlation matrix: understanding redundancies between variables*

Making correlations between descriptors considered two by two allows us to verify redundancies between variables. Spearman's correlation is advised, since it highlights increasing or decreasing monotonic relationships, unlike

Pearson's correlation, which only underlines relationships of a linear form. Besides, it constitutes a non-parametric approach that, unlike linear models, needs no applicability conditions. The function *panel.cor.spearman()*, found on a forum of the R software, is much more powerful than the function *pairs()* on its own. The use of the function *panel.cor.spearman()* in the argument upper.panel of the function *pairs()* displays results that are easier to interpret.

```
panel.cor.spearman <- function(x, y, digits=2, prefix="", cex.cor)
{
    usr <- par("usr"); on.exit(par(usr))
    par(usr = c(0, 1, 0, 1))
    r <- (cor(x, y))
    txt <- format(c(r, 0.123456789), digits=digits)[1]
    txt <- paste(prefix, txt, sep="")
    if(missing(cex.cor)) cex <- 0.8/strwidth(txt)
    test <- cor.test(x,y,method="spearman")
    Signif <- symnum(test$p.value, corr = FALSE, na = FALSE,
            cutpoints = c(0, 0.001, 0.01, 0.05, 0.1, 1),
            symbols = c("***", "**", "*", ".", " "))
    text(0.5, 0.5, txt, cex = cex * abs(r))
    text(.8, .8, Signif, cex=cex, col=2)
}
```

pairs (CP, lower.panel=panel.smooth, upper.panel=panel.cor.spearman)

We can observe strong positive correlations ($P < 0.001$) between the N/P ratio and nitrates (rho = 0.87), and between nitrites and turbidity (rho = 0.76) (Figure 1.3). A strong negative correlation ($P = 0.01$) can also be observed between phosphate and luminosity (rho = -0.68). Optical depth is significantly and negatively correlated to temperature (rho = -0.54) and nitrites (rho = -0.50). Thus, the variables put aside by Escoufier's method are redundant in relation to those that have been considered. This explains why the omission of these variables does not cause the global database to lose a lot of information. While interpreting data, it is advisable to keep in mind these redundancies between variables in order to better interpret the results. In some cases, it is preferable to consider all of them, since out of two redundant variables, one of them may turn out to be more explanatory than the other in relation to a database that needs to be explained (e.g. biological communities related to physical--chemical parameters).

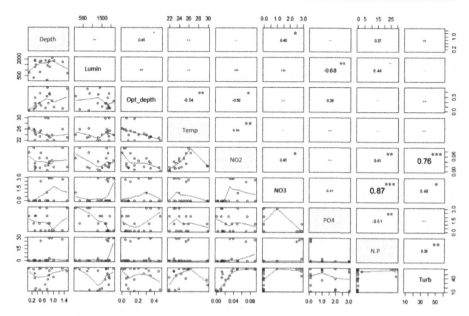

Figure 1.3. *Spearman's correlation matrix between all the descriptors taken two by two. The size of the correlation coefficients is shown proportionally to their importance (stars represent the levels of importance according to conventional criteria). The graphs crossed two by two are added on the left. The variables kept with Escoufier's method are shown in yellow (see section 1.3.2). For a color version of this figure, see www.iste.co.uk/david/data.zip*

1.4. Eliminating rare species

1.4.1. Why is it relevant?

When a flora or fauna database presents several different taxonomic units (in our case, 68 taxa), limiting the number of species to consider might be wise. A powerful analysis of the data might thus be carried out without being too influenced by species that in the end are not particularly substantial from an ecological point of view: their presence is actually severely limited, their abundances are low and these species, therefore, will not be important in characterizing the communities studied.

To this end, some authors use techniques that allow them to eliminate rare species, i.e. those that exhibit values of 0 in several stations. However, some of these species may be significantly abundant in the stations where they are inventoried and may even dominate the community in one or a few stations. The choice made and the ways used to eliminate rare species represent an essential, if complex, aspect: we must consider the number of stations where the species is present, as well as its relative presence in relation to the other species at these stations. We have to find a good balance between these two criteria, which is something that often depends on the data set used. Testing several methods on the same data set may allow us to get a better grasp of the role played by these two criteria. Here, we will use two techniques, out of the numerous methods available, for our phytoplankton data.

1.4.2. The median value method

The median value method has been used by Umaña-Villalobos [UMA 10]. According to this author, a rare species presents abundances that are smaller than the median values of all the species present.

The medians of abundances are calculated for each taxon for all stations. The database is transposed beforehand, so that the taxa are arranged into rows and the stations into columns. Then, a box plot is created in order to observe the data (Figure 1.4).

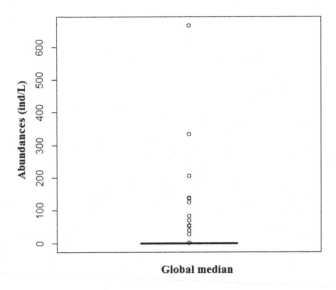

Figure 1.4. *Dispersion of the medians of 68 phytoplankton taxa calculated for 19 stations. The median of the medians is equal to 0*

medianglob <- apply(phyto, 2, median) # *Vector with the medians by taxon*
boxplot (medianglob, xlab="Global median", ylab="Abundances (ind/L)")

All the outliers in the box plots correspond to the taxa whose medians are greater than the global median (median of the medians by taxon), which in this case is equal to 0. Thus, they will correspond to non-rare species.

subset(t(phyto),medianglob>median(medianglob))->phyto2 # *Selection of the taxa whose medians are greater than the median of the medians*

dim(phyto2)
phytoS<-t(phyto2)

With this method, 15 taxa are considered as non-rare, with medians greater than the global median equal to 0. Therefore, this method would leave out 53 taxa.

1.4.3. *The abundance sorting method*

The "abundance sorting" method or ASM was adapted from Ibanez *et al.* [IBA 93] in the package [*pastecs*] by Grosjean and Ibanez [GRO 14]. It simultaneously takes into consideration the number of 0 present in the base for a given species, as well as the rare species (several 0) that are, however, abundant in some stations. A coefficient f, which ranges from 0 to 1, allows us to adjust the weight given to the frequency of the 0 values for a species ($f = 0$ if only this criterion is considered) and to the abundance of the species expressed in log.

Grosjean and Ibanez [GRO 04] propose to use a value of $f = 0.2$ to group species in four classes: (A) abundant species with few null values, (B) species with many null values but locally abundant, (C) averagely abundant species (more than 50-60% of null values) with relatively low abundance, and (D) insignificant and scarcely abundant species with a large number of null values. The descriptor c (implying the value of f chosen) takes into consideration the percentages of values not equal to 0 and the log-transformed abundances. A first extraction of the rare species with f = 0.2, followed by a second extraction from the result obtained with f = 1, allows us to exclude scarcely abundant species with large numbers of null values.

In the data set used here, "rare species" will be those that are obtained at stage C and D. In our case, we used f = 0.4 as a threshold, since it takes into consideration the abundance of the taxa with more precision and allows us to illustrate what we obtain at stage A and B (impossible in our data set with the recommended threshold of f = 0.2). These two stages correspond to the taxa that will be kept.

The abundance sorting method (ASM) can be used through the package [*pastecs*] with the function *abund()* by specifying the data base (taxa arranged into columns and stations in rows) and the value of f. The function *summary()* of the result phy.abd will list in descending order the importance of the taxa in the analysis: for each taxon, (i) the percentage relative to the most abundant species (expressed in log, e.g. *Nitzchia* being in this case the most abundant taxa in the database) and (ii) the percentage of values not equal to 0 for this taxon (e.g. *Nitzchia* being present in 100% of the stations). The function *plot()* enables us to represent graphically these results (in red and blue, respectively) as well as the curve of the cumulative differences between these two curves (in black), allowing us to determine the thresholds between the several stages. The function *identify()* makes it possible to establish the thresholds. Two thresholds are interesting: the boundary between stage A and B, and the one between B and C. The former threshold corresponds to the lowest point of the black curve: we can see the stage A taxa on the left. The end of the plateau corresponds to the second threshold: we can see the taxa of stages A and B on the left. Finally, the function *extract()* enables us then to retrieve the corresponding variables in the object phy.

library(pastecs)

phy.abd <- abund(phyto, f=0.4)

plot(phy.abd, dpos=c(40,100),xlab="Taxa",main="Method PTA f=0.4")

phy.abd$n <- identify(phy.abd)# *A small plateau is visible, click on the end of the*

plateau

Number of variables extracted: 23 on a total of 68

phy <- extract(phy.abd,phy.abd$n) #*New database with 23 taxa*

This method keeps 23 taxa (Figure 1.5) while the "median value" method advises us to consider 15 taxa. The method used, therefore, affects the number of taxa considered (or the number of rare taxa excluded). However, the results obtained are very coherent from one method to the other. The median method, which excludes the largest number of taxa, considers the same taxa as ASM.

Afterward, the results obtained with the median value method, which considers 15 species, will be kept. This considerably reduces the number of flora descriptors to consider in a multivariate analysis, making this number smaller than the number of stations (19) and thus allowing the applicability of certain analyses.

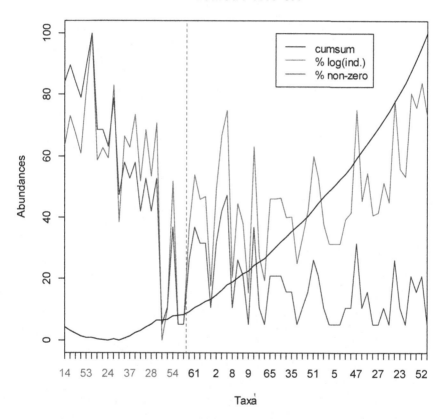

Method PTA f=0.4

Figure 1.5. *Abundance sorting method (ASM) for a threshold of 0.4. Non-rare taxa are shown according to their appearance numbers in the initial database on the right of the dotted red line. For a color version of this figure, see www.iste.co.uk/ david/data.zip*

2

Preliminary Treatment of the Data Set

The goal of this section is to illustrate how to calculate species richness and diversity indices (see section 2.1), how and why to transform the data set (see section 2.2) and how to choose the most suitable association measure in keeping with the scientific goals set beforehand (see section 2.3).

2.1. Abundances, species richness and species diversity

In this section, we will show how to calculate total abundance, taxonomic richness and the most common diversity indices by determining their specificity in relation to the complete phytoplankton database ("phyto", 68 taxa). All the results will be summarized in a table that presents for each station (row) the results of the indices at the 19 different stations.

Indices<-as.data.frame(matrix (c(0),nrow=nrow(phyto), ncol=2)) # *creation of the table*

rownames(Indices) <- rownames(phyto) #*Change the name of the rows*

colnames(Indices) <-c("Ab", "RS") #*Provide the name of the columns*

View(Indices)

2.1.1. Total abundances

Total abundances are calculated with the function *apply()* and the argument "sum".

Ab<-apply(phyto, 1, sum); Indices$Ab<-round(Ab,0) # ***Total abundances rounded to units***

barplot(Indices$Ab, col="grey",border = "black", names.arg = toupper(names(Ab)),
    main = "Abundances", xlab = "Stations", ylab = "Abundances (ind/L)",
    axes = TRUE) *#Barplot representing the total abundances per station (Figure 2.1)*

2.1.2. Species richness

Species richness corresponds to the **number of species found in an element** (here, a station). We should refer to taxonomic richness, since we are dealing with phytoplankton taxa rather than species. Despite this difference in the level of determination, the richness results may be compared between stations, since it has been measured on the same taxonomic level and by the same person for all stations. Taxonomic or species richness can be found with the function *specnumber()*, which belongs to the library [vegan]. The results are then stored in the "Indices" tables.

```
library(vegan)
RS<-specnumber(phyto); Indices$RS<-RS #Taxonomic richness
```

barplot(Indices$RS, col="grey",border = "black", names.arg = toupper(names(RS)),
    main = "Taxonomic richness", xlab = "Stations", ylab = "Taxonomic
    richness", axes = TRUE) *#Barplot of the taxonomic richness per station (Figure 2.1(B))*

Other species richness indices are also being put forward to overcome sampling-related problems, especially those concerning sample sizes, such as **Magalef's index** and **Odum's index** [GRA 05].

RSMargalef<-(RS-1)/log(Ab) ; Indices$RSMargalef<-RSMargalef

RSOdum<-RS/log(Ab) ; Indices$RSOdum<-RSOdum; head(Indices)

View(Indices) *#Two new variables have been added to the table*

t(Indices[,c("RSMargalef", "RSOdum")])->div1

barplot(div1, beside = TRUE, horiz = FALSE, col = c("grey", "white"), names.arg
    = toupper(colnames(div1)), legend.text = TRUE)

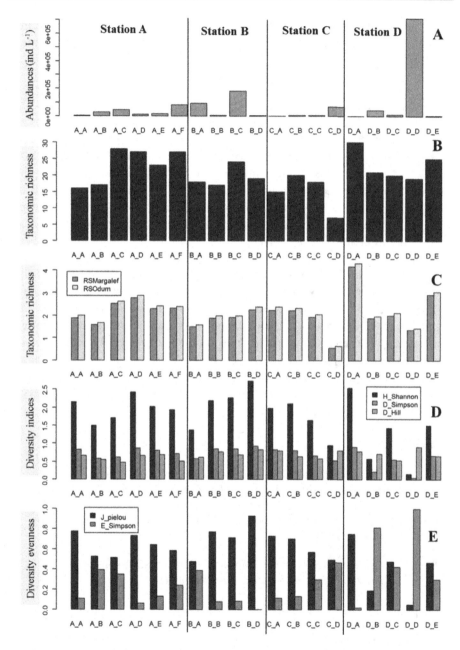

Figure 2.1. *Abundances A); taxonomic richness B); Margalef and Odum's richness C); Shannon, Simpson and Hill's diversity D); and Piélou and Simpson's evenness or equitability E) for the 19 stations*

In this case, the between-stations progressions are very similar for the classic taxonomic richness index, Margalef's index and Odum's index (Figure 2.1(C)).

2.1.3. *Diversity indices*

Species diversity indices take into consideration not only **species richness**, but also the **evenness between species**, namely the distribution throughout the stations of the number of individuals belonging to the species. For example, in a group represented by 10 species, if only one of them is highly abundant in relation to the others, the group is not very diversified, whereas if the species are well balanced in terms of abundance, the group will be more diversified.

The **Shannon–Wiener index and Simpson's index** [GRA 05] are the two most used indices. While Shannon's index is sensitive to the variations in importance of the rarest species, Simpson's index is sensitive to the variations in importance of the most abundant ones. These indices are often associated with evenness indices to interpret them correctly. This gives more weight to rare species, if we consider **Piélou's evenness,** or to the most abundant ones with **Simpson's evenness** [GRA 05]. **Hill's index** combines these two indices and provides a slightly more synthetic view [GRA 05]. Formulations of the indices can be found in all ecology books.

library(vegan)

H_Shannon<-diversity(phyto,index = "shannon"); Indices$H_Shannon<-
    H_Shannon # *Shannon-Wiener index, expressed as H*

J_pielou<- H_Shannon/log(RS); Indices$J_pielou<-J_pielou # *Piélou's evenness,*
    expressed as J

D_Simpson<-diversity(phyto,index = "simpson") ; Indices$D_Simpson<-
    D_Simpson # *Simpson's index, expressed as D*

E_Simpson<-(1-D_Simpson-min(1-D_Simpson))/(max(1-D_Simpson)-min(1-
    D_Simpson)); Indices$E_Simpson<-E_Simpson # *Simpson's evenness,*
    expressed as E

D_Hill<-diversity(phyto,index = "invsimpson")/exp(H_Shannon);Indices$D_Hill<-
   D_Hill # *Hill's index*

View(Indices); names(Indices*)* # *The new variables have been added to the table*

Graphic representations

t(Indices[,c("H_Shannon", "D_Simpson", "D_Hill")])->div2

barplot(div2, beside = TRUE, horiz = FALSE, col = c("black", "dark grey",

   "white"), names.arg = toupper(colnames(div2)), legend.text = TRUE)

t(Indices[,c("J_pielou", "E_Simpson")])->div3 #Figure 2.1(D)

barplot(div3, beside = TRUE, horiz = FALSE, col = c("black", "grey"), names.arg =

   toupper(colnames(div3)), legend.text = TRUE) #Figure 2.1(E)

Abundances fluctuate widely from one station to another: the total abundances of phytoplankton are very low at all stations, except for one station in marsh D (Figure 2.1(A)). Taxonomic richness and the diversity and evenness indices vary more from one station to another within the same marsh than they do between marshes, except for station C, where abundances and richness are globally lower (Figure 2.1). The Shannon, Simpson and Hill diversity indices are very similar in terms of evolution within the marshes, except for stations D_B and D_D: Hill's index yields a more important diversity in relation to the Shannon and Simpson indices. The evolution of Shannon's evenness closely resembles that of the diversity indices, whereas Simpson's evenness evolves in the opposite direction. The diversity of the stations seems to be sensitive to the variations in the abundances of rare species, instead of reacting to either taxonomic richness or the dominant species in the marshes of the Charente-Maritime department during the summer.

For the rest of the analyses, the "Indices" database, which includes total abundance, richness and diversity indices, has been combined with the "bio" database, which lists phytoplankton biomasses, production and productivity.

bio<-cbind(Indices,bio); head(bio)

2.2. Transformations

Data must be transformed in most cases before any numerical type of treatment (see section 1.2.2) [LEG 98] (1) to overcome the problems related to heterogeneous units involving variation ranges of descriptors without any ecological consequences (see section 2.2.1); (2) to limit the very strong fluctuations of the variation range of units in homogeneous data, thus giving less weight to the variables highly present in the analyses (e.g. taxa, see section 2.2.2.1); (3) to degrade a data set in order to make it homogeneous in terms of sampling technique (see section 2.2.2.2); (4) to quantify qualitative data (see section 2.2.3); and (5) to normalize data (see section 2.2.4). However, these transformations are not trivial, and they involve distortions of raw data that may be more or less strong (see sections 2.2.1.2 and 2.2.2.1).

2.2.1. *Quantitative data expressed with heterogeneous units (e.g. chemical physical data)*

2.2.1.1. *Standardization/standard score*

Standardization or standard score transformation is the solution adopted to overcome the problem of data expressed in heterogeneous units. Standard score transformations, which involve dividing by the standard deviation of each descriptor, ensure that the descriptors are expressed in the same unit, i.e. standard deviation, while centering implies the mean value to be 0.

For example, in the chemical–physical database, we can standardize data by using the function *scale()*.

```
CPtrf<-scale(CP2); head(CPtrf)
```

2.2.1.2. *Consequences in terms of data distortion*

This transformation, just like any other, has certain consequences in relation to the distance between points projected into a multidimensional space. As an example, the group of stations is projected into a two-dimensional space (optical depth in y and depth in x; Table I.1), with raw data (Figure 2.2(A)) and standardized data (Figure 2.2(B)) in the same variation range of y and x values.

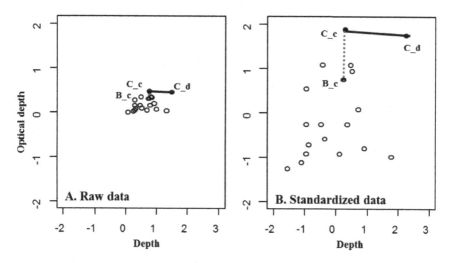

Figure 2.2. *Consequences of standardization on the distance between points in a plane formed by two variables. Example of the optical depth–depth relationship. The same variation ranges have been used for the x and y axes*

The distances between the points C_d, C_c and B_c can be considered as examples to observe the differences in projection between raw and transformed data. Let us notice that not only the distances C_d-C_c (solid line) and C_c-B_c (dashed line) have changed, but also the proportions between them have evolved. The ratio C_c-B_c/C_d-C_c is much more important in transformed, rather than raw, data: point B_c is less close to C_c by comparison with C_d in the transformed data.

The projection of the points has thus been distorted due to the standardization. The optical depth range was quite limited and, therefore, it has been widened by the transformation so that it can be compared to the depth range. The objective of the standardization is, thus, to express all the descriptors in the same unit: standard deviation.

This transformation is required in order to avoid giving too much weight to the variables that present high variation ranges because of their units. However, it is important to be aware of the transformation performed in relation to further analyses and the interpretation of the results.

2.2.2. Quantitative data expressed with homogeneous units (e.g. flora and fauna)

2.2.2.1. Reducing variation ranges

Once the descriptors are all expressed with the same unit (for example phytoplankton abundance in cells L^{-1}), simple transformations such as **logarithm** (log(x + 1), function *log1p()*), **square root** (function *sqrt()*) or **double square root** (function *sqrt(sqrt())*) are in this case the most appropriate. In this kind of transformation, variation ranges are reduced and the zero that represents the absence of a species remains a null value. The distortions created by these three transformations are not the same (Figure 2.3).

We consider here the projection of the stations onto the plane formed by the abundances of *Nitzchia*, in *x*, and *Cyclotella*, in *y*. Raw data (Figure 2.3(A)) allows us to see that maximum abundances are 100 times greater for *Nitzchia*. Station D_d is characterized by the highest abundances of *Nitzchia* and the absence of *Cyclotella*, whereas station A_f presents the highest abundances of *Cyclotella* and very low abundances of *Nitzchia*. Apart from these two extreme values, the abundances of the two taxa in other stations are much lower (except for *Cyclotella* in stations B_a and B_c). All the stations are superposed around 0 and the graphic representation is strongly affected by the two extreme values, which will consequently greatly influence the assessment of the differences between the stations.

The gap between the different stations, as well as the proportions between two respective gaps on the plane (for example 1_f-B_a in green and A-f-B-c in red), changes gradually with the square root (Figure 2.3(B)), double square root (Figure 2.3(C)) and log(x + 1) (Figure 2.3(D)) transformations, with more marked differences and more reduced variation ranges for the last transformation. Thus, the distortions of raw data will be more accentuated with a log(x + 1) transformation than with a double square root transformation. Similarly, the distortions will be more marked with a double square root transformation than with a square root transformation. Consequently, choosing between these three kinds of transformation depends on the width of the variation ranges in the data. The constant goal is to be influenced as little as possible by the extreme values, while also minimizing the distortions as much as we can.

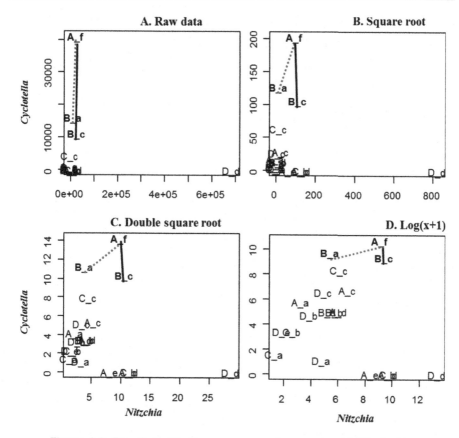

Figure 2.3. *Raw Data (A). Consequences of a square root (B), double square root (C) and log(x + 1) (D) transformation on the distance between points in a plane formed by the abundances of two taxa, Cyclotella and Nitzchia. For a color version of this figure, see www.iste.co.uk/david/data.zip*

These three transformations are performed here on the simplified "phytoS" database, which only contains the 15 non-rare taxa. The preliminary graphical representation of the data dispersion with a boxplot function allows us to choose the most suitable transformation for the data set considered. Ideally, we should reduce the variation ranges without overwriting them too much.

phytosqrt<-sqrt(phytoS) ; boxplot(phytosqrt) *#square root transformation and graphical representation of the dispersion*

phytodsqrt<-sqrt(sqrt(phytoS)) ;   boxplot(phytodsqrt)   *#double square root transformation*

phytolog<-log1p(phytoS), boxplot(phytolog) # *log-transformation*

2.2.2.2. Degradation of quantitative data into binary or qualitative data

2.2.2.2.1. Degradation into binary data

In some cases, we should consider only the presence or absence of species due to sampling anomalies or other aspects (section 1.2.2). Data are therefore recoded as 0/1: 0 for the absence of the species and 1 for values not equal to 0. This action can be performed on the raw database with the function *decostand()*, which belongs to the library [vegan]. Let us consider the example of the simplified "phytoS" base, which only includes 15 phytoplankton taxa.

library(vegan)

phytoPA<-decostand(phytoS,method="pa"); View(phytoPA)

2.2.2.2.2. Degradation into qualitative data

Quantitative data can be degraded into qualitative variables for simplification purposes. Let us consider the example of the variable "surface of the drainage basin" ("DB") in the database "environmental factors". We will code that all the DBs that are bigger than 6,000 ha are large DBs expressed as "3", those smaller than 2,000 ha as small DBs expressed as "1", while the other medium ones are expressed as "2". This variable will be later transformed into an ordinal qualitative variable (factor variable).

fext$DB[fext$DB<2000]<-1;fext$DB[fext$DB>2000&fext$DB<6000]

    <-fext$DB[fext$DB>6000]<-3

fext$DB<-as.factor(fext$DB) # *Transformation into a factor variable*

2.2.3. *Qualitative data (e.g. environmental factors)*

Occasionally, it is useful to transform qualitative data into quantitative data in order to be able to incorporate them into multivariate analyses. **This is the principle of a complete disjunctive table.** Each qualitative variable takes different modalities. In a complete disjunctive table, a qualitative variable will be replaced by a number of binary variables proportional to the number of modalities that the qualitative descriptor in question can take.

As an example, the "predominant land use on the DB" descriptor (designated by "use"; Figure 2.4) is a nominal qualitative variable with three modalities: "Farmland", "Grassland" or "Urban".

	Qualitative variable	New binary variables		
Code	Land use	Land use. Farmland	Land use. Grassland	Land use. Urban
Di_a	Grassland	0	1	0
Di_b	Farmland	1	0	0
Di_c	Farmland	1	0	0
Di_d	Farmland	1	0	0
De_b	Grassland	0	1	0
De_c	Urban	0	0	1
De_d	Grassland	0	1	0
Re_b	Urban	0	0	1
Re_c	Urban	0	0	1

Figure 2.4. *Transformation of a qualitative variable into new binary variables, one for each modality of the descriptor. Example on the "land use" nominal qualitative variable*

In a complete disjunctive table, this variable will be transformed into three binary variables: "use.Farmland", "use.Grassland" and "use.Urban". Each variable will, in each row, take a value of 1 if the "Use" qualitative descriptor has taken this modality, or 0 if this is not the case. A complete disjunctive table can be created from a table of qualitative variables with the function *acm.disjunctive()*, which belongs to the library [ade4].

library(ade4)

fext.disj<-acm.disjonctif(fext simple interlign) *#All variables must be recognized as factors. Change the kind of variable for each one with as.factor() function*

head(fext.disj)

2.2.4. *Attempting to normalize data*

Transformations can also be used to ensure that the data concerning a variable follow the normal distribution or another theoretical distribution required for the application of certain numerical analyses (e.g. environmental variables before a principal component analysis, or abundances of species before the application of a linear model). Standardizations (see section 2.2.1.1) may in some cases normalize data, while logarithmic transformations may normalize abundance data (see section 2.2.2.1). However, most of the time we should not expect any miracles.

2.3. Association measures and matrices

The general term *"association"* describes any kind of measures or coefficient used to quantify similarities or differences between objects or descriptors [LEG 98]. There are several types of association measures, and their use depends on the analysis mode and the kind of variable.

2.3.1. *How to choose our measure?*

In **direct mode**, when we describe similarities between objects, **similarity, difference** and **distance measures** are used, whereas in **indirect mode**, when we describe how descriptors depend on each other, **dependence coefficients** are preferred: covariance or correlation. The formulas that allow us to calculate these measures and the resulting association matrices will not be explained here, so that we can concentrate mainly on the elements that play a significant role in how these coefficients and matrices are chosen in relation to the type of data and the goals of the study (Table 2.1):

– *Similarity measures* describe the degree of similarity between objects or descriptors. They vary between 0 (total difference) and 1 (total similarity) or between 0 and 100 according to the software.

– *Dissimilarity measures* describe the degree of dissimilarity between objects or descriptors. They vary between 1 (total difference) and 0 (total similarity) or between 100 and 0 according to the software. The dissimilarity is equal to 1 or 100 minus the similarity for the same kind of coefficient.

Mode	Type of data	Double zeroes as similarity criterion	Coefficients	Characteristics	R functions
DIRECT (association between objects)	BINARY	Yes	*Simple concordance dissimilarity*	S=0.25 => 25% of coabsence or copresence	library(ade4); dist.binary(x, method=2)
		No	*Jaccard's dissimilarity*	S=0.25 => 25% of copresence	library(ade4); dist.binary(x, method=1)
			Sorensen's dissimilarity	Double weight to copresence	library(ade4); dist.binary(x, method=5)
	QUANTITATIVE	Yes	*Euclidean distance*	Shortest geometrical distance, on standardized data	dist(x, method="euclidean")
		No	*Bray-Curtis dissimilarity*	Abundance data, raw or transformed, abundance scaling, often used	library(vegan); vegdist(x, method="bray")
			Chi-squared distance	More importance to the profile than to raw abundances, sensitive to rare species	library(vegan); dist(decostand(x,"chi.square"), method="euclidean")
INDIRECT (association between descriptors)	BINARY	Yes	*Simple concordance dissimilaruty*	S=0.25 => 25% of coabsence or copresence	library(ade4); dist.binary(x, method=2)
		No	*Jaccard's dissimilarity*	S=0.25 => 25% of copresence	library(ade4); dist.binary(x, method=1)
			Sorensen's dissimilarity	Double weight to copresence	library(ade4); dist.binary(x, method=5)
	QUANTITATIVE	Yes	*Pearson's linear correlation*	Sensitive to the applicability conditions of linear models; assumes straight-line relationships	1-abs(as.dist(cor(x, method="pearson")))
			Kendall or Spearman's monotonic correlation	No applicability conditions; based on ranks; does not assume any form of monotonic relationship	1-abs(as.dist(cor(x, method="kendall"))) OR 1-abs(as.dist(cor(x, method="spearman")))
		No	*Chi-squared distance*	More importance to the profile than to raw abundances; sensitive to rare species	library(vegan); dist(decostand(x,"chi.square"), method="euclidean")

Table 2.1. *Key table allowing the choice of association measure for the most classically used ones as well as the functions for their applications in R*

– *Distances* describe the degree of dissimilarity between objects or descriptors, and they vary between 0 (totally similar objects or descriptors) and infinity (totally dissimilar). However, distances can also be normalized, so that they vary between 0 and 1. Normalized or unnormalized distances may also be transformed into similarities [LEG 98].

– *Correlation coefficients* describe the degree of similarity between descriptors and vary between –1 and 1.

Association measures must be chosen according to their properties in relation to the data set in question. This choice is made (Table 2.2) in relation to:

– indirect or direct mode;

– the kind of data (for example binary and quantitative);

– how much weight we wish to give (or not give) to "double zeroes". The association measures that take into consideration a *"double zero"* as a similarity criterion are called *symmetric*, whereas those who discard it are named *asymmetric*. Thus, we will use a symmetric coefficient for chemical–physical parameters that consider zero as a value, whereas we will more commonly use an asymmetric coefficient for phytoplankton abundances for which zero represents the absence of a species;

– certain properties that we may wish to consider or not when processing our data, e.g. giving more or less weight to copresence, profiles, differences in abundances, etc.

2.3.2. *Direct mode*

2.3.2.1. *The case of symmetric coefficients: significant "double zeroes"*

2.3.2.1.1. Association matrix of the stations based on binary data

The environmental factors of the marsh database appear now as a **matrix of binary data** called "fext.disj" because of a transformation through a complete disjunctive table. In this data table, it is important to take into consideration **double zeroes and no data transformation** is required since we are dealing with binary data. Thus, we will use the **dissimilarity coefficient of simple concordance** (Table 2.1; [SOK 58]) to create an association matrix with the function ***dist.binary()*** belonging to the library [ade4].

library(ade4) ; MATfext<-dist.binary(fext.disj,method=2)

MATfext   *#Displaying the dissimilarity matrix of simple concordance*

2.3.2.1.2. Association matrix of the stations based on quantitative data

The chemical–physical factors of the marsh database appear as a **matrix of quantitative data** that have been standardized to overcome the heterogeneous unit-related problem (i.e. "CPtrf" object). In this data table, it is important to consider **double zeroes**, since zeroes represent ordinary values and do not imply that data are missing. The use of the Euclidean distance on the transformed data is thus advised in order to create the association matrix (Table 2.1).

MATphys<-dist(CPtrf, method = "euclidean")

MATphys

2.3.2.2. The case of asymmetric coefficients: discarded "double zeroes"

The fifteen most common phytoplankton taxa may be determined in two ways: either with a degraded presence/absence matrix (i.e. "phytoPA" object) or with a quantitative matrix transformed beforehand to avoid giving too much weight to the taxa that may be particularly abundant in certain stations (i.e. "phytoS" object). In both cases, *a double zero* represents the

coabsence of a taxon in two stations and **will not be considered** as a criterion of similarity between the stations.

2.3.2.2.1. Association matrix of the stations based on binary data

The Jaccard distance ([JAC 08]; Table 2.1) will be used for the presence/absence matrix.

library(ade4)

matphyPA<-dist.binary(phytoPA,method=1)

2.3.2.2.2. Association matrix of the stations based on quantitative data

As for the taxon abundance matrix, we will use the Bray–Curtis dissimilarity coefficient (1957) on a log(x + 1) transformation.

library(vegan)

MATphyto<-vegdist(log1p(phytoS), method="bray")

2.3.3. Indirect mode

In indirect mode, the analysis is based on the relationships between variables. A Spearman correlation matrix based on the complete chemical–physical database can be built, for example, to establish the relationships between chemical–physical variables ("CP", Table 2.1). This correlation matrix is then transformed into a distance matrix by considering the following formula: $D = 1 - | C |$, where D is the distance and C represents the Spearman correlation coefficient for each pair of variables. Thus, the relationships highlighted do not take into consideration the sign of the relationship: two variables with a low distance present a strong correlation that may be positive or negative. The *as.dist()* function allows us to transform the correlation matrix into a similarity matrix that can be used later on, whereas the function *abs()* enables us to obtain the absolute value of the Spearman coefficient.

```
DISTphys<-1-abs(as.dist(cor(CP,method="spearman")))
```

3

Structure as Groups of Objects/Variables

Cluster analysis may be considered as a non-parametric approach, since it does not involve any applicability condition. Thus, it can be applied to both "small" and "large" data sets. However, on its own it will only provide the structures of objects or descriptors, without giving any information about what generates the groups observed (i.e. which descriptors account for the groups of elements? Which relationships are there among descriptors within one group?). The first part of this chapter introduces the most commonly used types of cluster analysis in environmental science (see section 3.1), while the second part presents the analyses that may allow us to obtain information about the descriptors that have generated the groups obtained in direct mode to complete the former analysis (see section 3.2).

3.1. The most used types of cluster analysis

These kinds of analyses involve **looking for discontinuities in the whole of the data**, namely grouping objects (direct mode) OR descriptors (indirect mode) similar enough to be able to be combined in the same group. There are several methods or algorithms, involving different **decision criteria**, that can compare elements two by two and create these groups [LEG 98].

Clustering defines the multidimensional analysis itself, which aims to partition a set of elements (objects OR descriptors). A **partition** is a division of the set into subsets, so that each element belongs to only one subset. A **cluster** is the result of the analysis: it may be constituted by a single partition or several hierarchized partitions.

3.1.1. *How to choose a clustering algorithm?*

Clustering algorithms are quite varied and the resulting partitions may be different in relation to the decision criteria used by one or the other. There are several kinds of methods. We distinguish between:

– **agglomerative clustering** (elements are considered separately at first, and then linked step by step by adopting an increasingly broader bottom-up tree approach) or **divisive clustering** (top-down approach, considering all the elements at the beginning);

– **hierarchical clustering** (subsets of elements that are classed in progressively larger groups) or **non-hierarchical clustering** (a unique partition that optimizes intragroup homogeneity);

– **exclusive clustering** (an element belongs to only one group) or **fuzzy clustering** (each element has a probability of belonging to a certain group after the final partition).

Clustering	Method	Type	Advantages/Drawbacks	Recommended for…
Hierarchical	Simple relationships through pairs of elements	Single-linkage	Rarely optimal; chain effect hindering well-defined groups	Good complement to analyses highlighting gradients; rarely used
Hierarchical	Simple relationships through pairs of elements	Complete-linkage	Rarely optimal; compact groups with a low level of similarity	To simply increase the contrast among groups; rarely used
Hierarchical	Combined relationships through pairs of elements	Unweighted means (UMGMA)	Fusion of groups when similarity reaches the mean among groups	Random or systematic sampling
Hierarchical	Combined relationships through pairs of elements	Weighted means (WPGMA)	Likewise, with adjustment according to the size of the groups	Other types of sampling
Hierarchical	Minimizes the sum of intragroup squares	Ward's method	Applicability conditions of principal component analysis	Complementary to principal component analyses
Non-hierarchical	Minimizes the sum of intra-group squares	K-means clustering	Number of groups given at the beginning and different optimal numbers of possible groups	Partition into the target number of groups
Fuzzy	Minimizes the sum of intra-group squares	C-means clustering	Number of groups given at the beginning and different optimal numbers of possible groups	Provides the probabilities for each element of belonging to each group

Table 3.1. *Key table that allows us to choose a clustering method: only the most classically used have been listed*

Only the most common methods used in environmental science will be described here. We will emphasize their advantages, drawbacks and applicability conditions in relation to the type of data (Table 3.1). These methods involve polythetic – all the variables considered to partition elements, i.e. descriptors, are used on each level – and simultaneous clustering (the process is carried out only once based on well-defined decision criteria).

3.1.2. Hierarchical clustering

Hierarchical clustering is the most common and intuitive type of clustering. It can be applied in direct or indirect mode.

3.1.2.1. The most used algorithms

To illustrate the formation of hierarchical clusters, we will build them from a six-station reduced matrix (i.e. six stations of marsh A) in the simplified and log-transformed "phytoplankton" database.

– Single-linkage hierarchical clustering [LUK 51]

This is the simplest type of clustering, as its name suggests. The dissimilarities obtained in the Bray–Curtis dissimilarity matrix are ranked in increasing order: the lower the dissimilarities, the closer the stations (Figure 3.1(A)). Here, the closest stations are "c" and "d" and they converge on the tree at a dissimilarity of 0.104. Assigning another station to this group only requires that the station is the closest one to one of the elements of the group formed. This is the case for station "f", which presents a dissimilarity of 0.186 in relation to station "c" in the group formed by "c" and "d". Stations "b" and then "e" follow the same principle. We have to wait until the dissimilarity reaches 0.431 before station "a" may be close to one of the other four stations.

– Complete-linkage hierarchical clustering [SØR 48]

This type of clustering differs from the single-linkage algorithm in that an element will join a group of elements when it is linked to the most distant member of the group (Figure 3.1(B)). Stations "c" and "d" are the closest, with a dissimilarity of 0.104. However, a group formed by stations "e" and "f" is created for a dissimilarity of 0.226, before station "b" is linked to the group "c-d" (0.238). The group "e-f" will then link itself to the group "c-d-b" when the dissimilarity is equal to 0.341 and "a" to the other stations at a dissimilarity equal to 0.703.

A. Single-linkage hierarchical clustering

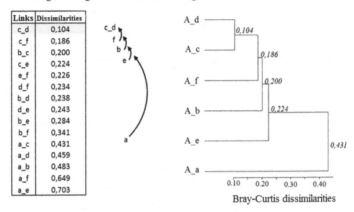

B. Complete-linkage hierarchical clustering

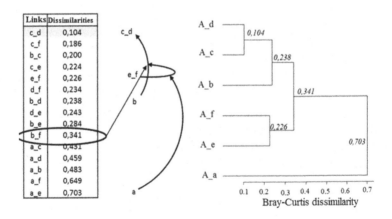

Figure 3.1. *Schematic representation of the construction of hierarchical trees based on single-linkage A) and complete-linkage B) algorithms*

– Mean-linkage hierarchical clustering [SNE 73]

There are several types of mean-linkage hierarchical clustering. They are based on calculations concerning mean similarities between elements or groups, rather than the number of links between elements. For example, the Unweighted Pair Group Method with Arithmetic Mean (UPGMA) algorithm deals with unweighted means, whereas the Weighted Pair Group Method with Arithmetic Mean (WPGMA) algorithm deals with weighted means.

– **Ward's method** [WAR 63]

Ward's method combines elements in order to minimize the sum of squares of the distances of each group from the centroid. In other words, its goal is to minimize intragroup variance. This method is very suitable for Euclidean distance matrices. Therefore, it should be avoided when we are dealing with asymmetric coefficients (for example Bray–Curtis dissimilarities), i.e. while studying the structure of biological communities.

Two hierarchical clusters are proposed to reach two different goals as examples of use in Table 3.1:

– **To determine groups of stations in the database of the phytoplankton communities (direct mode).** A hierarchical cluster is built from the Bray–Curtis dissimilarity matrix ("MATphyto") created with the simplified and log-transformed phytoplankton taxa database.

– **To determine relationships between chemical–physical descriptors (direct mode).** A hierarchical cluster is applied to the distance matrix based on the Spearman correlation matrix between chemical–physical variables ("DISTphys").

3.1.2.2. *Creating clusters and displaying classifications*

Clusters are created with the function ***hclust()*** and the argument *method* that enables the choice of clustering algorithm. Then, hierarchical trees can be displayed with the function ***plot()***. The argument *hang* = −1 allows us to range all the stations on the *x*-axis, so that the graph is more elegant.

3.1.2.2.1. Direct mode: establishing groups of stations according to the phytoplankton communities

Three algorithms out of those presented here can be used to establish groups of stations in the database of the phytoplankton communities: single-linkage, complete-linkage and UPGMA. They are applied to the "MATphyto" Bray–Curtis dissimilarity matrix (Figure 3.2). Ward's method should be avoided when we are using assymetric coefficients such as the Bray–Curtis dissimilarity.

#*Application of clustering algorithms*

cah.single.phy<-hclust(MATphyto, method="single") # *Single-linkage*

cah.complete.phy<-hclust(MATphyto, method="complete") # *Complete-linkage*

cah.UPGMA.phy<-hclust(MATphyto, method="average") # *UPGMA*

#*Graphic representation of the associated classifications*

plot(cah.single.phy,hang=-1, main="Single linkage", xlab="Stations", ylab=" Bray-Curtis dissimilarity") # *Single-linkage*

plot(cah.complete.phy,hang=-1, main="Complete linkage", xlab="Stations", ylab="Dissimilarités de Bray-Curtis") #*Complete-linkage*

plot(cah.UPGMA.phy,hang=-1, main="UPGMA", xlab="Stations", ylab=" Bray-Curtis dissimilarity") # *UPGMA*

Groups of stations can be determined by cutting the tree at a constant value of the association coefficient (here, a dissimilarity), such as 0.3 or 0.6. These three methods do not yield the same groups of stations due to the differences in the building algorithms.

For the "single-linkage" algorithm, a cut at 0.35 would yield four groups, three of which would include a single variable (C_d, A_a and C_a), while the fourth would gather the other stations (Figure 3.2). The "single-linkage" mathematical algorithm creates chain-reaction effects: stations converge toward the tree one after the other, following the principle by which it was built. Thus, this method makes it easier to highlight gradients rather than groups. Therefore, it is very often supplementary to unconstrained ordination. Otherwise, the "complete-linkage" algorithm regulates more strictly how the stations join the tree and, consequently, it allows us to underscore more distinct groups (Figure 3.2). This is the case for a dissimilarity of 0.6, which allows us to determine three groups, each of which includes at least two stations. Most of the time, these groups are too compact and do not actually correspond to a real ecological context. The UPGMA algorithm represents a good compromise between these two algorithms, but it requires a type of sampling that has to be either completely random or thoroughly systematic (Figure 3.2; [LEG 98]).

3.1.2.2.2. Indirect mode: relationships between chemical–physical variables

The four algorithms – single-linkage, complete-linkage, UPGMA and Ward's method – can be applied here to the DISTphys distance matrix, built from Spearman correlations between variables (see section 2.3.3), to establish relationships between chemical–physical variables.

cah.single.CP<-hclust(DISTphys, method="single") ; plot(cah.single.CP, hang=-1, main=" Single linkage ", xlab="Stations", ylab="Distances") # *Single - linkage*

cah.complete.CP<-hclust(DISTphys, method="complete"); plot(cah.complete.CP, hang=-1, main="Complete linkage", xlab="Stations", ylab="Distances") #*Complete-linkage*

cah.UPGMA.CP<-hclust(DISTphys, method="average"); plot(cah.UPGMA.PC, hang=-1, main="UPGMA", xlab="Stations", ylab="Distances") # *UPGMA linkage*

cah.ward.CP<-hclust(DISTphys, method="ward"); plot(cah.ward.CP, hang=-1, main="Ward", xlab="Stations", ylab="Distances") # *Ward's method*

As was the case for the analysis carried out in direct mode, these four algorithms do not yield the same results (Figure 3.2). Therefore, the questions arising are as follows: (1) which algorithm is most suitable for the data processed (see section 3.1.2.3) and (2) for which distance or dissimilarity value should the classification be cut to obtain relevant groups of elements (see section 3.1.2.4).

3.1.2.3. *Choosing the most suitable clustering algorithm with cophenetic distances*

The cophenetic distance between two elements of a hierarchical tree represents the length that separates one of the elements from the closest node that links it to another element [SNE 73]. Therefore, it measures the concrete distances between two elements in a specific tree. A cophenetic matrix is a matrix that lists all the distances between the pairs of elements in a tree. The comparison between the base association matrix and the cophenetic matrix of the tree resulting from one of the algorithms allows us to take into consideration the actual associations between elements entailed by the algorithm used. The graphical representation of the cophenetic distances in relation to the association coefficients, all pairs of objects taken together, and the Pearson correlation coefficient allows us to then measure the relationship between the two matrices. This will enable us to assess the extent to which an algorithm can correctly represent the association between pairs of elements, and, consequently, to compare several algorithms in order to choose the one that closely represents the ecological situation [BOR 11].

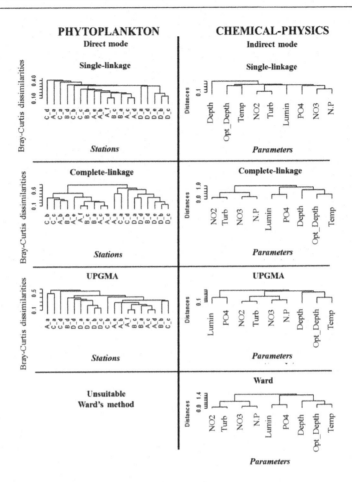

Figure 3.2. *Graphical representations of the four hierarchical agglomerations most used in environmental science to group stations based on their phytoplankton communities (15 taxa) in direct mode (on the left) and chemical–physical descriptors in indirect mode (on the right): single-linkage, complete-linkage, UPGMA and Ward. Ward's method is not suitable for the treatment of abundances data as an assymetric coefficient has been applied*

The cophenetic matrix can be calculated with the function *cophenetic()* based on the tree built. The function *cor()* allows us to then calculate Pearson's correlation between the cophenetic and the association matrix. These operations are carried out to determine which algorithm, out of the three employed, is the most appropriate when we want to establish groups of stations in the database of the phytoplankton communities (Figure 3.2).

coph.single<-cophenetic(cah.single.phy) #*"Single-linkage" cophenetic matrix*

coph.complete<-cophenetic(cah.complete.phy) #*"Complete-linkage" cophenetic matrix*

coph.average <-cophenetic(cah.UPGMA.phy) # *"Mean linkage" cophenetic matrix*

cor(MATphyto, coph.single) # *Pearson's correlation coefficient*

cor(MATphyto, coph.complete)

cor(MATphyto, coph.average)

The best Pearson correlation is obtained for the UPGMA algorithm (r = 0.8432). For the following analyses, we will thus use the UPGMA algorithm, to build the hierarchical classification, since it is the one that closely represents the dissimilarities between pairs of elements in this case.

3.1.2.4. *Choosing which group number to consider: the similarity profile method (SIMPROF test)*

SIMilarity PROFile analysis (SIMPROF test) was developed by Clarke *et al.* [CLA 08]. It is a permutation test that employs a null hypothesis that should be tested to identify structures within the ecological communities of groups of elements that differ significantly concerning the descriptors database. It is based on an association matrix (Figure 3.3). It can also be applied to environmental parameters (for example chemical–physical variables).

The null hypothesis corresponds to the absence of a multivariate structure in the descriptors database (Figure 3.3). The SIMPROF test examines whether the similarities observed between the elements are more or less marked than those obtained randomly. The similarities calculated for each pair of objects in the association matrix are arranged in increasing order: the shape of the curve of the similarities in relation to the similarity ranks represents the similarity profile of the initial database (points in bold in Figure 3.3). To test the null hypothesis, data are randomly permuted many times over (i.e. 999 or 9,999) in the initial base matrix, the association matrix recalculated each time, and then the random similarity profiles obtained. The set of randomly obtained similarity profiles is averaged, and the mean curve corresponds to the theoretical profile if the null hypothesis is true (i.e. lack of group structure). The $\prod$ permutation test statistic corresponds to the distance between the actual profile and the mean theoretical profile under H0. This same distance is calculated for all the theoretical profiles. A frequency plot of the distances from the theoretical profiles is drawn, and it looks like an unimodal curve. The

∏ statistic is compared with the theoretical frequency diagram. The more dissimilar it is from this diagram, the more H0 is rejected. In this respect, this test is very similar to a parametric test in which the test statistics are compared with a normal distribution. We obtain a P-value as a classical test (P-value < 0.05, H0 rejected) and, therefore, we can observe a significant structure in the data (Figure 3.3). This operation is carried out again for each node of the hierarchical tree according to the algorithm we have chosen (for example the UPGMA algorithm is used for the 15 phytoplankton taxa database) in order to identify significant groups in our example.

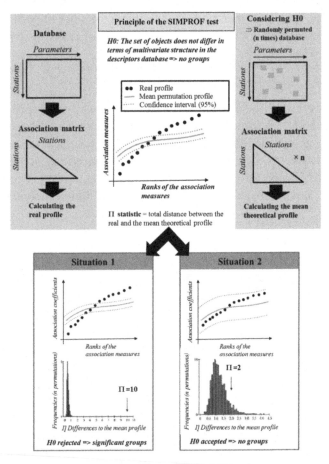

Figure 3.3. *Schematic representation of the appropriate statistical approach in the analysis of the SIMilarity PROFile test (SIMPROF)*

SIMPROF analysis can be carried out with the function *simprof()*, which belongs to the [clustsig] library. It requires us to specify the transformed database in the argument *data*, the association coefficient in the argument *method.cluster* and the clustering algorithm in the argument *method.distance*. The corresponding *plot()* function displays the various groups in different colors.

library(clustsig)

simprof.UPGMA <- simprof(data=log1p(phytoS), num.expected=10000, num.simulated=9999,method.cluster="average",method.distance="braycurt is",sample.orientation="row") #*These calculations may take some time (since the same operation must be carried out many times over)*

simprof.UPGMA ; summary(simprof.UPGMA)

simprof.UPGMA$significantclusters # *5 significant groups*

The SIMPROF test underlines here five significantly different communities. We can see how the results are slightly different from those yielded by a classical cut (for a constant value of dissimilarity), since this test is carried out for each node of the tree (Figure 3.4). With the SIMPROF test, stations A_c and C_a are put in the same group, whereas they belong to two different one-station groups in the classic five-group cut. Similarly, with the SIMPROF test, station C_b is isolated in a group aside, while it was joined to a larger group when we performed the classical cut.

In R, the *cutree()* function does not involve the vector that indicates membership to the groups obtained with the SIMPROF approach, since it cuts classically at a constant level of dissimilarity, as is the case for a tree built with the *hclust()* function. Thus, this must be done by hand.

simprof.phyto<-c(1,5,5,5,5,5,5,5,5,5,5,3,1,4,5,2,3,3,3,3)

The groups obtained may be different for each SIMPROF procedure, since the results are based on theoretical distributions estimated with random permutations of the source data matrix each time. If the results vary too much each time, it is advisable to increase the number of random permutations performed to establish the theoretical distribution, so that this distribution can become more stable (the *num.expected* and *num.simulated* arguments go, for example, from 1,000 and 9,999 to 100,000 and 999,999, respectively). These calculations will be longer, but the results will be more reproducible.

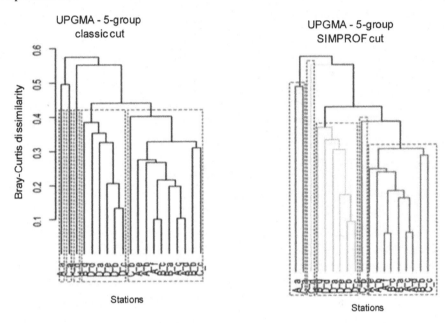

Figure 3.4. *Trees built by a UPGMA hierarchical agglomeration on a Bray–Curtis dissimilarities matrix created on the log-transformed database of the 15 phytoplankton taxa. The SIMPROF analysis (on the right) highlights five distinct groups of stations based on phytoplankton communities. The results differ slightly from those of a five-group cut obtained by means of a classic method with a constant level of dissimilarities (on the left). For a color version of this figure, see www.iste.co.uk/david/data.zip*

3.1.3. *k-Means non-hierarchical clustering*

The *k*-means non-hierarchical method employs the local structures of data to create groups by identifying high-density regions in the data [JAI 88]. The number of groups must be established at the beginning. This is a type of divisive non-hierarchical clustering that operates by minimizing the intragroup variance of raw data. Since the minimized variance is a sum of squares of Euclidean distances, it is therefore more likely to be used for data where double zeroes are important. As the number of groups must be established at the beginning, it is then wise to optimize the number of groups before the analysis.

This analysis is carried out on the "bio" database, which contains 12 diversity index and phytoplankton production variables and whose double zeroes are important. The number of variables has been reduced beforehand with Escoufier's method in order to eliminate redundant variables that do not provide a lot of information. In "bio_simp", seven variables have been considered: the Shannon–Weiner, Simpson and Margalef diversity indices; taxonomic richness; taxonomic abundances; and phytoplankton production and productivity. Afterward, the database is standardized and, later on, a Euclidean distance matrix is calculated.

bio_simp<-bio[,c(5,7,3,2,1,11,12)]; names(bio_simp); BIO.trf<-scale(bio_simp)

MATbio<-dist(BIO.trf,"euc")

In order to choose the optimal number of groups, we will try several *k*-means partitions with different groups and observe the silhouette plots obtained. These plots allow us to determine whether stations are properly arranged in each group. The width of the silhouette varies between −1 and 1. The higher the value is, the more properly the station is arranged in the group. If the width is negative, the station in question is arranged badly. The method put forward is based on the mean silhouette of all the elements for different cuts (two-group cuts, three-group cuts, etc.). It takes into consideration that the maximum mean width corresponds to the best cut and, therefore, to the best number of groups obtainable. The next loop allows us to draw the silhouette graphs (function *silhouette()* in the library [cluster]) for partitions from two to 10 groups, following [BOR 11].

```
library(cluster); par(mfrow=c(3,3))
for(i in c(2:10)) {
        bio.kmeans<-kmeans(BIO.trf, centers=i, nstart=100)
        plot(silhouette(bio.kmeans$cluster,MATbio),main="Silhouette - k means")
}
```

Ideally, we should find a limited number of groups (three to seven groups), with a minimum of groups including only one station and a maximum of stations well represented in each group. Here, this is the case for a four-group *k*-means partition that isolates station 18, but whose other groups gather between three and nine well-represented stations.

bio.kmeans<-kmeans(BIO.trf, centers=4, nstart=100) *#Specify the number of groups in the argument* **centers**

bio.grpe<-bio.kmeans$cluster; bio.grpe *#Assigning stations to the group*

3.1.4. *Fuzzy c-means clustering*

Together with the previous methods, which partition elements within groups that do not overlap, there is a number of clustering methods called "fuzzy", which can be both hierarchical and non-hierarchical [BOR 11]. These methods are based on the principle that some elements may be shared by several groups in the final partition.

This is the case for *c*-means clustering, which, in terms of principle and application conditions, closely resembles the *k*-means non-hierarchical clustering. However, it yields the probability that each element has of belonging to the different groups (the sum is equal to 1). According to the principle of fuzzy clustering, an element that is very likely to be part of one of the groups and very unlikely to be part of any other group clearly belongs to the group in question.

c-Means clustering is applied to the "bio" simplified database and to the resulting "MATbio" Euclidean distance matrix. It is carried out with the function *fanny()*, which belongs to the library [cluster]. As was the case for *k*-means non-hierarchical clustering, it is necessary to establish the number of groups. Let us choose four groups, which seems the best compromise for the most similar non-fuzzy method in terms of algorithms (*k*-means).

k<-4

library(cluster)

bio.fuz<-fanny(MATbio,k=k,memb.exp=1.5)

summary(bio.fuz)

biofuz.m<-round(bio.fuz$membership,3)#*Probability of belonging to the groups*

biofuz.g<-bio.fuz$clustering #*Fixed group membership*

biofuz.g

Stations like D_c and D_d are very likely to belong to a specific group (G1 and G4, respectively), whereas others are shared between several groups: A_a and A_c. These probabilities represent a bonus when it comes to interpreting a badly classed element, since they allow us to identify the several groups that share this element and to determine transitional elements between well-defined groups.

3.2. Information on the descriptors that generate the groups obtained

The drawback of these methods is that they only carry out partitions of elements, without giving us any information about the descriptors that have generated the groups, especially in direct mode.

3.2.1. Simple graphical representation

Some graphical representations allow us to characterize the groups in relation to the descriptors used for clustering. Let us consider, for example, the identification of groups of stations in the database of the phytoplankton communities. We have previously established five groups of stations thanks to the SIMPROF test (see section 3.1.2.4). The following graphical map gives us an idea of the distribution of each taxon by element. Abundances are averaged by station. The darker the color of a given station is, the higher the abundances of this taxon are in relation to the other taxa. This graphical map requires the function *vegemite()*, which belongs to the [vegan] library [BOR 11].

library(cluster); library(vegan); library(RColorBrewer)

finalclust.phy<-reorder.hclust(cah.UPGMA.phy,MATphyto) #*Rearrangement of the stations according to the constraints of the cluster analysis*

dend<-as.dendrogram(finalclust.phy)

or<-vegemite(log1p(phytoS), finalclust.phy, scale="Hill")

heatmap(t(log1p(phytoS)[rev(or$species)]), Rowv=NA, Colv=dend,
    col=c("white",brewer.pal (5, "Greens")), margin=c(4,4), ylab="taxa (means
    weighted by sites)", xlab="Stations")

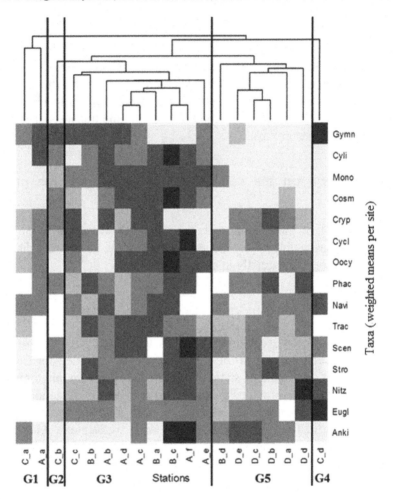

Figure 3.5. *Graphical map of the abundances compositions of the species in relation to the groups of stations according to the UPGMA algorithm applied to the log-transformed abundances. The abundances are weighted per station. The groups obtained with the SIMPROF test are represented*

The group G4, which contains only one element (C_d), is characterized by very high abundances of Euglena and *Gymnodinium* (Figure 3.5) while the group G5 is marked by very low abundances of the *Gymnodinium*,

Cylindrotheca, *Monoraphidium* and *Cosmarium* taxa; and the group G3 is characterized by fairly strong abundances of all taxa. Systematic or even strong abundances of *Gymnodinium*, *Cryptomonas*, *Oocystis* and *Navicula* are noticeable for G1. G2 presents strong abundances of *Gymnodinium*, *Cylindrotheca*, *Cosmarium*, *Cryptomonas*, *Phacus* and *Scenedesmus* in relation to the other taxa.

3.2.2. Analysis of variance

One-factor analysis of variance (ANOVA) allows us to explain a quantitative variable by means of a qualitative one. In this sense, it allows us **to analyze descriptors that present significant differences between groups**, e.g. the groups of stations that present different phytoplankton communities. ANOVA presents a parametric approach (classical ANOVA), non-parametric approaches (Kruskal–Wallis test and ANAlysis of SIMilarities (ANOSIM)), and an intermediate approach with less constraints in terms of applicability conditions (permutational ANOVA) than classical ANOVA. Section 5.2.1.7 describes in detail the application of these four general types of ANOVA as well as the verification of the relative applicability conditions.

3.2.3. Determining indicator species

Once groups of stations or dates have been created from databases concerning biological communities, it may seem interesting to determine the indicator species of each group identified. Here, we put forward two methods: considering an indicator species as a species specific to a given group and species fidelity to all the stations/dates of this group (IndVal, see section 3.2.3.1) or identifying species that contribute to the dissimilarity between groups (SIMPER, see section 3.2.3.2).

3.2.3.1. Species specific to one group and "faithful" to its stations ("IndVal")

The IndVal ("Species Indicator values") method was developed by Dufrêne and Legendre [DUF 97] to identify species regarded as indicators of groups, namely species that are specific and "faithful" to a given group. By specificity, we mean a species characterized by how it is only present in a given group. By fidelity, we mean a species that is present in all the elements of this group. This method simultaneously combines mean abundances with their frequency of occurrence in the different groups. The highest IndVal (equal to 1) for a given group means that a species is only present in this group and in all the elements of this group. A threshold value of 0.25 corresponds to a species that is present in at least 50% of the elements of the group and/or that its relative abundances reach at least 50%. This method is classically applied to raw abundances, even if they have been transformed for the previous analysis.

This method can be applied with the function *indval()*, which belongs to the library [labdsv]. This function calculates the IndVals for each species and group, and the following script allows us to summarize in a table the maximum IndVal obtained by species together with its corresponding group (drawn from [BOR 11]. It also allows us to carry out a permutation test that can highlight significant IndVal indicator values.

Here, this method is applied to determine whether there are indicator species for the five groups of stations created with the SIMPROF method considering the phytoplankton communities (allocating stations to the groups obtained in the object simprof.phyto).

```
library(labdsv)

indval(phytoS,simprof.phyto)->IV;IV

gr<-IV$maxcls;gr #group of stations where each taxon has a max indval

iv<-IV$indcls;iv #group of stations with indval equal to the max indval

pv<-IV$pval;pv #corresponding p-values

tabIV<-data.frame(group=gr,indval=iv,pvalue=pv);tabIV #Table showing which
    groups the indicator species belong to and the corresponding p-value

tabIV[order(tabIV$pvalue),] # Sorting the IndVal according to the p-value
    obtained with the permutation tests
```

All taxa present IndVals greater than 0.25 for a group, which confirms that they are present in at least 50% of the elements of this group and/or that their relative abundances reach at least 50% in it. It is important to highlight that, in this example, not all the groups present indicator taxa. Moreover, only two taxa constitute significant indicators of group 5 according to the related permutation tests: *Scenedesmus* (IV = 0.82; *P*-value = 0.04) and *Monoraphidium* (IV = 0.66; *P*-value = 0.045).

These IndVal indicators could have been applied to the set of 68 taxa (i.e. by conserving the rare taxa).

3.2.3.2. *Similarity percentages (SIMPER)*

This analysis aims to determine how much each species contributes to the dissimilarity between stations [CLA 93]. It allows us to identify the most significant species in decreasing order, so that we can distinguish between the dissimilarity profiles observed between the types of a qualitative factor (i.e. membership in the groups highlighted). This analysis works on the Bray–Curtis dissimilarities database. Therefore, data must lend itself to this association coefficient (fauna or flora matrix). We consider in general those species that contribute to 75% of the dissimilarity between groups, taken two by two. This analysis is incremented with the function *simper()*, which belongs to the library [vegan].

The analysis needs at least two elements per group.

This type of analysis requires us to specify the transformed database as well as a qualitative variable that tells us which group the elements belong to. Here, it is applied to the simplified and log-transformed phytoplankton database to determine which taxa contribute to the dissimilarity between the groups highlighted by the SIMPROF. Groups 2 and 4, which present only one station per group, are first removed from the data set.

library(vegan)

fext$simprof.phyto<-as.factor(simprof.phyto) # *Addition of the column allocation of the stations to the 5 groups to the "environmental factors" data set*

phytoS[simprof.phyto!=2&simprof.phyto!=4,]->phy # *removing the stations corresponding to groups 2 and 4, which only present one station in the "phytoplankton" base* fext[simprof.phyto!=2&simprof.phyto!=4]->grpe # *removing the stations corresponding to groups 2 and 4, which only present one station in the "environmental factors" base*

sim<-simper(log1p(phytoS),grpe $simprof.phyto) # *For the 3 groups, we obtain 3 sub-tables corresponding to the taxa that contribute to the dissimilarity between groups (1 and 5, expressed as Contrast 1_5, 1 and 3, expressed as contrast 1_3, 5 and 3, expressed as contrast 5_3). Here, we provide only the results for table 1_5.*

#*As an example: comparison group 1 and group 5*
sim$'1_5'->tab1_5

tab1_5$average # *for the contribution to overall dissimilarity*

tab1_5$averall # *The overall between-group dissimilarity*

tab1_5$ava # *Average abundances for group 1*

tab1_5$avb # *Average abundances for group 2*

tab1_5$ord # *An index vector to order vectors by their contribution*

tab1_5ordcusum # *Ordered cumulative contribution*

Each subtable corresponds to a comparison between groups taken 2 by 2. Here we only interpret the subtable comparing group 1 and group 5. For each subtable, taxa are sorted in relation to the decreasing significance of their contribution to the dissimilarity between the two groups. The *cumsum* column lists the cumulative dissimilarity for each taxon added, whereas the av.a and av.b columns list the transformed mean abundances between the two groups considered (a for the former – in this case, group 1 – and b for the latter – in this case, group 5).

Ten taxa contribute to 75% of the dissimilarity between the stations belonging to group 1 and those of group 5. In decreasing order of significance, they are *Trachelomonas, Monoraphidium, Nitzchia,* Euglena, *Scenedesmus, Strombomonas, Cryptomonas, Oocystis, Cosmarium* and *Cyclotella*. All taxa are more abundant in group 5, except for *Cryptomonas*, which is more abundant in the first group. As for the other two tables, 10 taxa contribute to 75% of the dissimilarity between the stations of group 1 and those of group 3. In decreasing order of significance, they are as follows: *Cylindrotheca, Trachelomonas, Euglena, Gymnodinium, Ankitrodesmium, Strombomonas, Nitzchia, Navicula, Scenedesmus* and *Cosmarium*, with higher abundance values in group 3, except for *Cylindrotheca, Gymnodinium* and *Cosmarium*. Ten taxa contribute to 75% of the dissimilarity between the stations of group 3 and those of group 5. In decreasing order of significance, they are as follows: *Monoraphidium, Nitzchia, Cylindrotheca, Cosmarium, Cyclotella, Cryptomonas, Oocystis, Scenedesmus, Ankitrodesmium* and *Gymnodinium*, with higher abundances in group 5, except for *Cryptomonas* and *Ankitrodesmium*.

4

Structure as Gradients of Objects/Variables

A parametric approach (unconstrained ordination, see section 4.1), used for "large" data sets, has the advantage of providing on its own structures of objects and the descriptors that account for them, unlike a non-parametric approach (see non-metric multidimensional scaling (nMDS), section 4.2), suitable for "small" data sets, which only yields the structure of objects or descriptors in relation to the scientific goal determined beforehand. Besides, the axes of unconstrained ordination hierarchize the information (i.e. to determine the parameters that shape the structure of the elements in a predominant or secondary way), unlike a non-parametric approach, where the axes are meaningless. Thus, these two criteria make parametric approaches, through unconstrained ordination, the most powerful such kind of analysis.

4.1. Parametric alternative: unconstrained ordination

This aims to summarize the information included in a matrix with x variables or parameters, namely to **deduce the general trends that emerge from it following certain axes** [LEG 98]. The goal is to reduce the number of axes of a multidimensional matrix (i.e. each variable represents a dimension) to some axes that contain the main information. In order to illustrate the principle of this analysis in a simple way, let us consider the classic example of the camel, a three-dimensional (3D) object (Figure 4.1). Plane B would be more suitable than plane A if we wanted to reduce this object to a plane (two dimensions) that contains the main information. It actually represents the Bactrian camel in profile and allows us to distinguish the two humps, whereas plane A, which represents the animal frontally, may as well be showing a dromedary camel. Plane B provides more information than plane A for the same reduced number of axes.

Figure 4.1. *Reduction of the image of a 3D object to a plane containing the maximum amount of information*

The main advantage of this analysis is that it allows us to **hierarchize the information included in the source data matrix**: the first axis extracted contains the most important information, the second axis the important information once the first axis has been extracted and so on. All the axes are therefore independent. Objects and variables will be represented according to these axes and each of them will take on an ecological meaning.

Therefore, this analysis is more powerful than a multidimensional scaling analysis, where only the position of the elements (object or variable, in relation to the mode used) is available, and whose axes are ecologically meaningless (see section 4.2). **Be careful! This analysis can only be carried out in direct mode**: objects (in rows) and parameters (in columns) are processed in different ways and according to their type while pieces of information are provided for both objects (e.g. stations) and parameters.

Despite being tempting, this analysis presents **applicability conditions** that must be respected, which makes it less robust than multidimensional scaling. Thus, we refer to parametric multivariate analysis. Its mathematical principle is based on the fact that **all variables follow the normal distribution** (i.e. data multinormality), which is rarely the case! We could easily do without this condition if the database had 10 times more objects than variables. However, it is usually accepted for data sets that contain at least more objects than variables.

4.1.1. *Mathematical principle*

Its geometric principle is illustrated by the kind of unconstrained ordination that is most commonly used, i.e. principal component analysis (PCA) ([HOT 33]; Figure 4.2). This analysis uses the Euclidean distances between objects (the easiest type of distance to understand in geometric terms) to identify the factorial axes. Let us consider, for example, a matrix with three parameters – x, y and z – and 27 objects. Figure 4.2(A) shows the 27 objects in a 3D space (x–y–z). If data follows the normal distribution for the three variables x, y and z, then the objects are distributed according to a 3D ellipsoid (Figure 4.2(B)), namely a sphere elongated in relation to the multinormality of the variables considered. The center of the ellipsoid is the point of inertia of the point cloud. The length of the ellipsoid is the direction in which the point cloud stretches to its maximum and, consequently, where variance is the greatest: it is, thus, defined as the first principal component. This vector represents a simple linear combination of the variables x, y and z (Figure 4.2(C)). The second principal component goes through a plane perpendicular to the first component through the point of inertia of the ellipsoid (Figure 4.2(D)). This is the direction in which the cloud of the points projected on to this plane stretches as much as possible. It is also a linear combination of the variables x, y and z (Figure 4.2(E)). The plane formed by these two first components is called the principal plane (i.e. plane

B in the example of the camel shown in Figure 4.1). The first principal plane maximizes the distance between the points projected vertically on to this plane and the point of inertia of the point cloud (Figure 4.2(G)). According to the Pythagorean theorem, the vertical projection of the points called e_i is therefore minimal. This allows us to say that the principal plane is the one that passes nearest to all the points and, consequently, preserves as well as possible the position of the objects in the original space (Figure 4.2(H)).

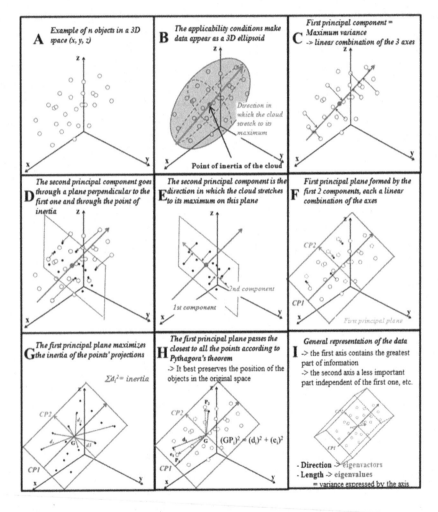

Figure 4.2. *Geometrical principle that governs how we obtain the factorial axes based on a 3D matrix while applying a principal component analysis. For a color version of this figure, see www.iste.co.uk/david/data.zip*

The third component can only go in one direction in the case of a 3D matrix, since it will be perpendicular to the two first components when passing through the point of inertia (Figure 4.2(I)). The first component includes the greatest amount of information, the second one a slightly smaller amount of information, which is different from the first one, etc. We refer to the independence of the information conveyed by the axes. The number of components that can be extracted will be equal to the number of initial variables. However, only the first few of them are used, since our aim is to summarize the information. The direction of the components (linear combinations of the variables x, y and z) is called "eigenvector" and their length "eigenvalue". An eigenvalue represents the inertia expressed by an axis, namely the amount of information it conveys. If we use the Euclidean distance or if we are carrying out PCA, the inertia corresponds to the variance.

4.1.2. Principal component analysis

4.1.2.1. When should we carry out PCA?

Since this kind of analysis is based on the Euclidean distance, double zeroes are important and should be considered in the initial database as a similarity criterion between two elements (e.g. chemical–physical variables, see section 1.2.5). Most of the time, data are heterogeneous in terms of units. Thus, it is important to standardize it to overcome the problems linked to differences concerning variation ranges (see section 2.2.1). When units are homogeneous, standardization requires some thought in relation to the goals of the study.

PCA is applied in **direct mode** to **quantitative or semi-quantitative data**. Following the mathematical principles involved in its application, we assume that if there are relationships between the parameters, they are of a **linear** kind.

Finally, as we have previously discussed, data should follow the normal distribution for all the parameters. However, a number of objects greater than the number of parameters can be accepted to overcome this problem.

4.1.2.2. Mathematical principle

PCA considers objects in the space of the parameters according to the Euclidean distance. Eigenvalues and eigenvectors can be obtained by working on the covariance matrix of the parameters (i.e. diagonalization). If we consider standardized data, as is often the case when we want to homogenize units, this operation is carried out on the correlation matrix (Figure 4.2). Each axis presents:

– an **eigenvalue** that represents the variance explained by the axis, i.e. the quantity of information conveyed by the axis;

– an **eigenvector**, i.e. the orientation of the axis in the space of the x parameters, which is nothing more than a linear combination of them.

The coordinates of the objects on each component are obtained by combining the data with the eigenvectors, while the coordinates of the parameters are found by combining the vectors with the eigenvalues (Figure 4.3). As for the graphical representation of the results, objects and parameters are not represented in the same space: **the graphs of the objects and those of the parameters are distinct.** The **correlation circle** for the cloud of the parameters can only be obtained once data are standardized at the beginning (Figure 4.3).

In R, there are several libraries that can be used to carry out a PCA ([vegan], [ade4], etc.). The **[FactoMineR] library**, with the function *PCA()*, remains the function that beginners can understand more easily, since it allows us to rapidly access quality indicators that it is important to consider in order to analyze the results of the PCA correctly. Here, the analysis will be carried out on the "chemical physical" database, while discarding those variables that do not convey a lot of information, as defined in section 1.3 (turbidity, luminosity, nitrates and optical depth). For now, the analysis will be carried out on the "CP" matrix by putting the discarded variables in the argument *quanti.sup*, which allows us to point out the supplementary variables that will be discussed later. These variables are not taken into consideration by the analysis, even if their projections are calculated. These variables are in the database CP. All the others parameters of CP are considered as "active" variables (considered for the analysis).

The argument *scale.unit* allows us to standardize the data. TRUE must be specified even if the data have been previously standardized (in case of standardized data, naturally!), so that the correlation circle can be represented in the graph of the parameters. It is not recommended to draw the graphs from the beginning (graph = FALSE), so that we will not be influenced by the results before the analysis of the quality indicators of the

PCA. The argument *ncp* points out the number of axes that should be considered for the results. For now, all axes are considered, i.e. as many as the parameters.

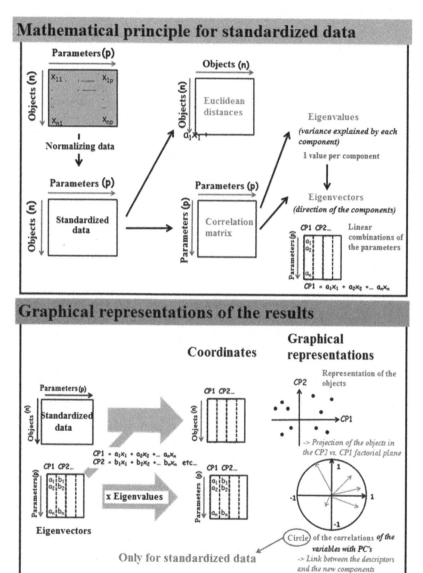

Figure 4.3. *Mathematical development of the implementation of a principal component analysis. For a color version of this figure, see www.iste.co.uk/david/data.zip*

library(FactoMineR)

names(CP) # *Initial chemical–physical database*

PCAphys<-PCA(scale(CP), scale.unit = TRUE, ncp = 5, quanti.sup = c(2,3,6,9), graph = FALSE) # *Redundant variables are classed as supplementary variables*

PCAphys Aphys #*To acceed to all the results available*

There are several tables that can account for the set of the results. We only have to call the object in which they are stored (i.e. PCAphys) followed by $ and the name of the results subset wanted. For example, PCAphys$eig lets us access the eigenvalues, PCAphysvarcoord the coordinates of the variables, etc. (see the available results by calling the object PCAphys).

4.1.2.3. Quality indicators

A) How many axes should we consider?

Three indices based on the eigenvalues help us determine the number of axes we should consider:

round(PCAphys$eig,2) # *Values rounded to two decimal places with the round() function*

	eigenvalue	percentage of variance	cumulative percentage of variance
comp 1	1.86	37.22	37.22
comp 2	1.34	26.73	63.95
comp 3	0.83	16.62	80.57
comp 4	0.72	14.31	94.88
comp 5	0.26	5.12	100.00

– eigenvalues greater than 1 which represents information more important than that provided by the mean of all variables. In this case, only the first two components;

– a percentage of explained variance equal to at least 50% (i.e. the summary of the information provided by the analysis will be greater than half of all). Here, the first two components already contain 64% of the information included in the database;

– a break in the gradient of the eigenvalues or of the variance explained by the several axes scree plot (the profiles of the two graphs are similar).

par(mfrow=c(1,2))

barplot(PCAphys$eig[,1], main="Eigenvalues scree plot", xlab = "Components", ylab="Eigenvalues")

barplot(PCAphys$eig[,2], main="Explained variances screeplot ", xlab = "Components", ylab="Explained variances")

We can observe a first break in the gradient between the third and the fourth axis (Figure 4.4). Thus, the first two indicators yield two axes, whereas the "eigenvalue scree plot" criterion yields three axes. Therefore, we will follow the common result and consider two axes: axis 1, explaining 37% of the total variance, and axis 2, explaining 27% of the total variance.

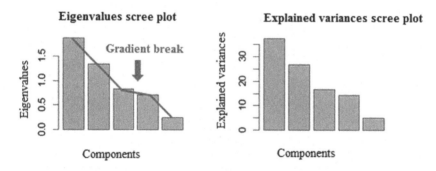

Figure 4.4. *Scree plot of the eigenvalues and the variance explained by the five axes of the principal component analysis carried out on the "chemical–physical" database*

Each axis considered must contribute some information. If we keep three axes and we realize later on that the third one does not convey any information (i.e. no variable is represented on this axis and no ecological meaning is found), keeping it is useless.

B) Which parameters can be interpreted?

We use the cosine^2 indicator to analyze the variables that can be interpreted on the first two axes. It represents the angle formed between the variable vector and the component in question. There are as many such angles as there are variables–axes pairs. For a given parameter, the sum of the cos^2, all components taken together, is equal to 1. The greater the cos^2 is (it varies between 0 and 1), the better the parameter is represented on the axis. The cos^2 on axes 1 and 2 will be analyzed independently from each other as well as on the principal plane (axis 1 + axis 2). If we arrange the cos^2 on the principal plane in decreasing order, we can highlight the variables that are better represented by the consideration of the first two axes.

quality <-round (data.frame (cos2axis1 = PCAphysvarcos2[,1],
        cos2axis2=PCAphysvarcos2[,2], cos2axis12 = PCAphysvarcos2[,1]
        +PCAphysvarcos2[,2]),2)

quality[rev(order(quality$cos2axis12)),] *#range by decreasing cos2 on the
        principal plane*

	cos2axis1	cos2axis2	cos2axis12
N.P	0.72	0.13	0.84
Temp	0.13	0.49	0.63
Depth	0.08	0.55	0.63
NO2	0.44	0.16	0.60
PO4	0.50	0.00	0.50

Only the variables N/P, depth, temperature and nitrites are well represented on the principal plane (cos^2 greater than the arbitrary value 0.6): the variables N/P and concentration of nitrites on axis 1, and the variables depth and temperature on axis 2. The phosphate variable, which is badly represented, cannot be interpreted on the principal plane.

The indicator pointing out how a parameter contributes to the construction of the axis allows us to give to the axis an ecological meaning. The sum of the contributions in the rows and columns is equal to 100.

contrib<-round(PCAphysvarcontrib[,1:2],2); contrib

	Dim.1	Dim.2
Depth	4.14	41.46
Temp	7.01	37.00
NO2	23.42	11.98
PO4	26.90	0.01
N.P	38.52	9.54

The N/P ratio on its own contributes 39% of the construction of the first axis, whereas depth and temperature contribute 41% and 37%, respectively, to the construction of the second axis. Note that some variables may be badly represented on one axis or another but correctly on the plane because they are dependent on both axes.

4.1.2.4. *Representation and interpretation of the results of the PCA*

Here are the script lines that allow us to find out the coordinates of the parameters and those of the stations, as well as the graphical representations themselves.

coordP<-round(PCAphysvarcoord[,1:2],2); coordP *#Coordinates on the correlation circle*

coordS<-round(PCAphysindcoord[,1:2],2); coordS *#Coordinates of the stations*

par(mfrow=c(2,2))

plot(PCAphys,choix="var",axes = c(1,2))

plot(PCAphys,choix="ind",axes = c(1,2))

On the correlation circle, the N/P ratio and the concentrations of nitrites are situated on the right part of the first axis: the more positive the coordinates of the stations on this axis, the higher the availability of nitrogen nutrients compared with phosphate nutrients (N/P ratio). On the second axis, temperature (positive value on the axis) and depth (negative value) face each other: the shallower the channel is, the higher the water temperature is. Thus, we can give to the two axes an ecological meaning: availability of nitrogen nutrients versus phosphate nutrients for axis 1 and temperature for axis 2 (Figure 4.5(A)).

As for the stations, it is quite tedious to read the results in Figure 4.5(B). In order to understand the structure of the stations, the *s.class()* function, which belongs to the library [ade4], allows us to group objects in the shape of ellipsoids according to the categories of a qualitative variable. In this example, this function is applied to the marsh (four categories: A–D; Figure 4.5(C)) and the kind of marsh (re-fed or unfed, Figure 4.5(D)).

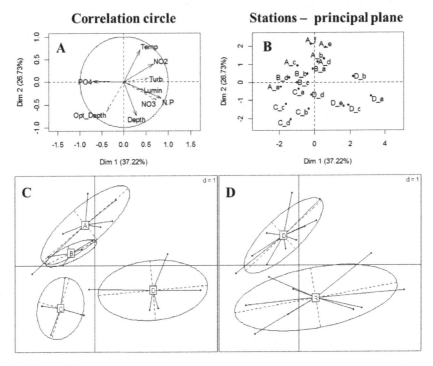

Figure 4.5. *Graphical representations of the principal component analysis applied to the "chemical–physical" database. A) Correlation circle. The supplementary variables are shown as dotted lines. B) Representation of the stations. C) Dispersion of the stations per marsh (A–D). D) Dispersion of the stations per type of marsh (drained unfed or re-fed)*

library(ade4)

s.class(PCAphysindcoord[,c(1,2)] , fac=fext$Station

s.class(PCAphysindcoord[,c(1,2)] , fac=fext$Type

Re-fed stations present higher availability of nitrogen nutrients and greater depth than unfed stations (Figures 4.5(C) and (D)). Marsh C is characterized by the lowest temperatures (Figure 4.5(C)). In the stations of unfed marshes (expressed as R), temperature dispersion is more important for station A than it is for station B (Figure 4.5(C)). In the case of the conservation of 3 axes, the same work had to be done on the plane formed by axis 3 vs axis 1 to interpret the third axis.

4.1.2.5. *Supplementary variables or individuals*

This analysis does not consider mathematically supplementary variables. However, their coordinates are projected onto the correlation circle (the dotted lines in Figure 4.5(A)). Here, the variables that had been discarded with Escoufier's method in section 1.3 (i.e. turbidity, luminosity, nitrates and optical depth) are shown as supplementary variables.

qualityvarsup <-round (data.frame (cos2axis1 = PCAphys$quanti.sup$cos2[,1], cos2axis2= PCAphys$quanti.sup$cos2 [,2], cos2axis12 = PCAphys$quanti.sup$cos2 [,1] + PCAphys$quanti.sup$cos2 [,2]),2)

qualityvarsup

	cos2axis1	cos2axis2	cos2axis12
Lumin	0.18	0.03	0.22
Opt_Depth	0.15	0.42	0.57
NO3	0.53	0.16	0.69
Turb	0.32	0.01	0.32

Only nitrates present a cos^2 value that guarantees a good representation on the principal plane ($cos^2 = 0.69$). If we consider the representation of the parameters on this plane, nitrates are well correlated to axis 1 and very close to the parameter N/P. This further confirms the ecological meaning given to axis 1.

Supplementary objects may also be represented with the argument *ind.sup* in the PCA function. Refer to the help section of the function by entering ?PCA.

4.1.3. Correspondence analysis

4.1.3.1. When should we carry out correspondence analysis?

Correspondence analysis (CA) is based on the χ^2 distance, which is **not affected by double zeroes** (i.e. phytoplankton data, see section 1.2.5; [ROU 67]). The main characteristic of the χ^2 distance is the **profile comparison**. If a variable varies from another variable by the same multiplicative factor for each object, these two variables will be considered as identical. For example, if, for each station, the abundances of the genus *Haslea* are 10 times greater than those of the genus *Caloneis*, CA points out that *Haslea* and *Caloneis* are identical, since they show the same profile between stations.

CA can only be carried out in **direct mode** and on **data that are homogeneous** in terms of units, **quantitative, semi-quantitative or binary**. Even though the extreme points are reduced by this type of distance, it is often necessary to transform data to avoid giving too much weight to variables with a wide variation range (see section 2.2.2.1).

As was the case for PCA, data should follow the normal distribution for all the parameters. However, in order to get around this condition, we can accept **a number of objects greater than the one of parameters**.

Finally, CA is based on the principle that if there is a relationship between the variables considered and some explanatory environmental variables, this relationship is unimodal. This condition is consistent with the fact that species show a relationship of this kind with environmental factors: an optimal response in terms of abundances in relation to fluctuations in temperature, turbidity, etc. (according to the principle of the species' ecological niche).

4.1.3.2. *Mathematical principle*

CA considers objects in the space of the parameters transformed into relative frequencies according to the χ^2 distance (Figure 4.6). The data table, which shows the relative frequencies of the objects in rows and the parameters in columns, is transformed into a table of standardized relative frequencies. Afterward, the covariance matrix of the parameters is calculated (Figure 4.6). Eigenvalues and eigenvectors are obtained by working on the covariance matrix of the parameters (i.e. diagonalization). As was the case for PCA, each axis presents:

1) an **eigenvalue** that represents the inertia explained by the axis, i.e. the amount of information conveyed by the axis;

2) an **eigenvector**, i.e. the orientation of the axis in the space of the x parameters, which is nothing more than a linear combination them.

Here, instead of talking of variance, we refer to the concept of explained inertia, since the χ^2 distance is a weighted distance.

The coordinates of the objects on each component are obtained by combining the data with the eigenvectors, whereas the coordinates of the parameters are obtained by combining vectors and eigenvalues (Figure 4.6). Unlike PCA, the results of the objects and those of the parameters are represented in the same space: we refer to the **duality of the representation space**. Here, the advantage is that we can make our interpretation in terms of influence: objects that are close to a parameter are influenced by it.

As was the case for PCA, the [FactoMineR] library, with the *CA()* function, remains the most intuitive type of library, since it allows us to more easily access the quality indicators that it is important to consider in order to analyze the results of CA correctly. Here, we will use the log-transformed database containing the abundances of the 15 phytoplankton taxa to illustrate how this analysis is carried out (phytoS).

The abundances are first of all log-transformed (log $(x + 1)$). As was the case for PCA, it is not recommended to draw graphs from the very beginning (graph = FALSE), to not be influenced by the results before the analysis of the quality indicators. The ncp argument specifies the number of axes we should consider for the results. As for now, we will consider all the axes, namely the same number of parameters minus 1.

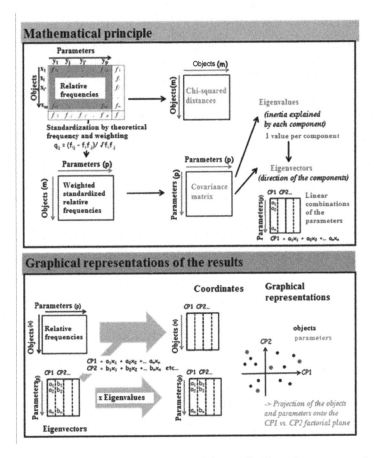

Figure 4.6. *Mathematical development of the application of a correspondence analysis. For a color version of this figure, see www.iste.co.uk/david/data.zip*

library(FactoMineR)

names(phytolog) # *Log-transformed database without the rare taxa (15 taxa)*

CA<-CA(log1p(phytoS), ncp = 14, graph = FALSE)

CA #All the results available

As was the case for PCA, there are several tables that can account for the set of the results. We only have to call the object in which they are stored (i.e. CA), followed by $ and the name of the results subset wanted.

4.1.3.3. Quality indicators

A) How many axes should we consider?

Only two out of the three indices for PCA help us determine the number of axes we should consider:

head(round(CA$eig,2))

	eigenvalue	percentage of variance	cumulative percentage of variance
dim 1	0.11	24.41	24.41
dim 2	0.10	20.88	45.28
dim 3	0.07	15.73	61.02
dim 4	0.05	10.45	71.47
dim 5	0.04	8.28	79.75
dim 6	0.03	5.51	85.27
dim 7	0.02	4.65	89.92
dim 8	0.02	3.79	93.71
dim 9	0.02	3.20	96.91
dim 10	0.01	1.12	98.03
dim 11	0.00	0.84	98.88
dim 12	0.00	0.63	99.50
dim 13	0.00	0.37	99.87
dim 14	0.00	0.13	100.00

– a percentage of explained inertia equal to at least 50%. Here, the first three components already contain 61% of the information of the database;

– a break in the gradient of the eigenvalues or of the inertia explained by the axes scree plot. Here, the change actually occurs between the fourth and the fifth axis, even though it starts taking place to a small degree between the third and the fourth one (Figure 4.7).

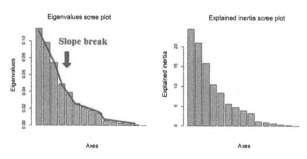

Figure 4.7. *Scree plot of the eigenvalues and the inertia explained by the 14 axes of the correspondence analysis carried out on the "phytoplankton" database*

```
par(mfrow=c(1,2));
```

```
barplot(CA$eig[,1], main="Eigenvalues scree plot", xlab="Axes",
ylab="Eigenvalues")
```

```
barplot(CA$eig[,2], main="Explained inertia scree plot", xlab="Axes",
ylab="Explained inertia")
```

Therefore, all the results seem to lead us to consider three axes: axis 1, which explains 24% of the total inertia; axis 2, which explains 21% of it; and axis 3, which explains 16% of it. As was the case for PCA, each axis considered must provide us with information. If we keep three axes and we realize later on that the third one does not provide us with any information (i.e. no variable is represented on this axis, no ecological meaning can be found), considering it would be useless.

B) Which parameters can be interpreted?

As was the case for PCA, we use the $\cos^2$ indicator to determine which variables (phytoplankton taxa) can be interpreted while we consider the first three axes of the CA.

```
quality <- round( data.frame( cos2axis1= CA$col$cos2[,1], cos2axis2 =
    CA$col$cos2[,2], cos2axis3 = CA$col$cos2[,3], cos2axis123 =
    CA$col$cos2[,1]+CA$col$cos2[,2]+CA$col$cos2[,3]),2)
```

```
quality[rev(order(quality$cos2axis123)), ] #The order() function allows us to sort
    in decreasing order according to the quality of the representation on the
    three axes added up (data not shown).
```

Only eight taxa out of the total 15 are well represented on the three axes considered, with a $\cos^2$ boundary of 0.6 (arbitrary value):

– On axis 1: *Cosmarium, Clyndrotheca, Monoraphidium*;

– On axis 2: *Gymnodinium, Ankistrodesmus*;

– On axis 3: *Cryptomonas, Nitzchia,* Euglena.

4.1.3.4. *Representation and interpretation of the results of the CA*

Here are the script lines that allow us to find out the coordinates of the taxa and of the stations, as well as the graphical representations themselves.

```
par(mfrow=c(3,2));

plot(CA, axes=c(1,2)) #Axis 2 versus Axis 1

plot(CA, axes=c(1,3)) #Axis 3 versus Axis 1

coordG<-round(CA$col$coord[,1:3],2); coordG #for species

coordGS<-round(CA$row$coord[,1:3],2); coordGS # for stations

library(ade4)

s.class(CA$row$coord[,c(1,2)] , fac=fext$Station)

s.class(CA$row$coord[,c(1,3)] , fac=fext$Station)

s.class(CA$row$coord[,c(1,2)] , fac=fext$Type)

s.class(CA$row$coord[,c(1,3)] , fac=fext$Type)
```

The three taxa *Cosmarium*, *Cylindrotheca* and *Monoraphidium* are typical of the positive part of axis 1 (Figure 4.8(A)). *Gymnodinium* and *Ankistrodesmus* are opposite each other on axis 2, the former among the positive values and the latter among the negative ones (Figure 4.8(A)). *Cryptomonas* is typical of the positive values of axis 3, unlike *Nizchia* and Euglena (Figure 4.8(B)). On axis 1, we can find the types of marshes opposite each other (Figures 4.8(E) and (F)): re-fed and unfed marshes are fairly influenced by *Cosmarium*, *Cylindrotheca* and *Monoraphidium*, unlike re-fed marshes. In re-fed marshes, the stations of marsh C are distributed distinctly along axis 1, unlike the other stations, and show negative values on axis 2. Therefore, they are influenced by *Ankistrodesmus*. The stations of marsh D have positive values on axis 2 and, consequently, they are influenced by *Gymnodinium* (Figure 4.8(C)). Stations C and A are substantially distributed on axis 3 (Figure 4.8(D)), influenced by *Cryptomonas*, *Nitzchia* and Euglena.

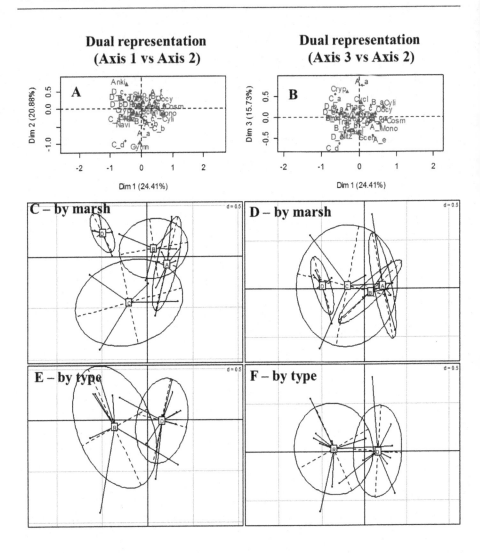

Figure 4.8. *Graphical representations of the CA applied to the "phytoplankton" database. Representation of the taxa (red) and stations (blue) on the second axis versus the first axis (A), and on the third axis versus the first one (B). Dispersion of the stations per marsh (A–D) on the second axis versus the first axis (C), and on the third axis versus the first one (D). (C) Dispersion of the stations per type of marsh (non-re-fed or re-fed) on the second axis versus the first axis (E), and on the third axis versus the first one (F)*

4.1.3.5. Special case: the horseshoe or Guttman effect

Species, being controlled by environmental factors, tend to show a unimodal distribution in relation to them. Thus, environmental gradients often present a succession of species (Figure 4.9(A); [LEG 98]). These successions of unimodal distributions can be represented mathematically by a horseshoe shape formed by stations and species, and they result in a dependence between axes 1 and 2 (Figure 4.9(B)).

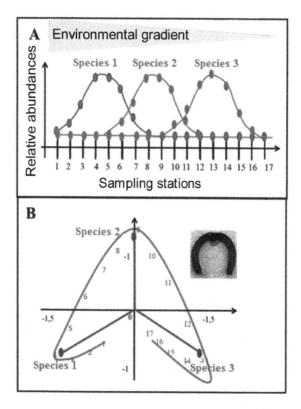

Figure 4.9. *Horseshoe effect: sequence of unimodal distributions of species along an environmental gradient (A), mathematical and geometrical consequences on the principal plane of the correspondence analysis (B)*

There are ways in which we can stop this dependence (i.e. detrended CA) and get rid of the horseshoe effect. However, they generate other types of mathematical bias, which is why they are being used less frequently [BOR 11]. The best option remains to interpret this arched form as the presence of an environmental gradient. Inertia is often very strong on the first axis (Figure 4.10), and this allows us to determine the environmental factor that originates this gradient with the analyses described in Chapter 5.

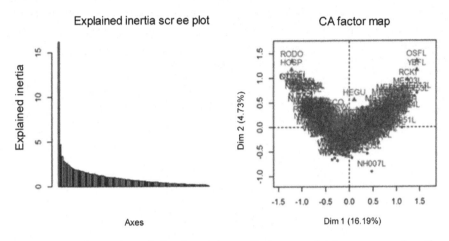

Figure 4.10. *Illustration of the horseshoe effect on a database presenting the presence/absence of 95 bird species inventoried in 208 lakes. The explained inertia scree plot shows well the predominance of the first axis. For a color version of this figure, see www.iste.co.uk/david/data.zip*

4.1.3.6. Supplementary variables and individuals

As was the case for PCA, supplementary variables or individuals that are not involved in the extraction of the axes during the analysis may be considered. This technique allows us to project the same variables used for the CA (i.e. taxa), but not environmental or chemical–physical factors, since they are not expressed in the same units of measurement. Therefore, the distance used by the CA does not correspond to the one that should be used for these factors.

4.1.4. Multiple correspondence analysis

4.1.4.1. When should we carry out multiple correspondence analysis?

Multiple correspondence analysis (MCA) is an **extension of CA** for **qualitative variables,** which are therefore subject to the same requirements (see section 4.1.3.1). Qualitative variables are transformed into binary variables with a **complete disjunctive table** beforehand, and then they are processed as they were in CA.

Here, we will carry out MCA on the "environmental factors" database. We do not need to create a complete disjunctive table beforehand, since the *MCA()* analysis   belonging to the library [FactoMineR] does that beforehand. For a more detailed explanation of the several stages shown below, readers should refer to the CA developed in section 4.1.3.

library(FactoMineR)

mca<-MCA(fext[,-8], ncp=5, graph=FALSE) #*All variables must be qualitatives. The simprof.phyto column must be removed if necessary*

mca

4.1.4.2. Quality indicators

A) How many axes should we consider?

The two indices that help us determine how many axes we should consider are as follows:

– a percentage of explained inertia equal to at least 50%, here 55% for the first two axes;

round(mca$eig,2)

	eigenvalue	percentage of variance	cumulative percentage of variance
dim 1	0.52	30.09	30.09
dim 2	0.42	24.72	54.82
dim 3	0.30	17.51	72.33
dim 4	0.18	10.69	83.01
dim 5	0.12	6.99	90.00
dim 6	0.09	5.35	95.36
dim 7	0.05	2.96	98.31
dim 8	0.03	1.69	100.00

– a break in the gradient of the eigenvalues or of the explained inertia scree plot; here, between the second and the third axis (Figure 4.11).

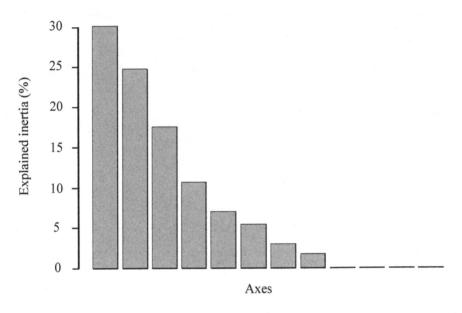

Figure 4.11. *Scree plot of the inertia explained by the nine axes of the MCA carried out on the "environmental factors" database*

```
par(mfrow=c(1,1))
```

```
barplot(mca$eig[,2], main="Scree plot of explained inertia",
xlab="Components",ylab="Explained inertia")
```

Therefore, all the results seem to lead us to consider two axes: axis 1, which accounts for 30% of the total inertia and axis 2, which accounts for 25% of it.

B) Which parameters can be interpreted?

```
qualitymca<-round(data.frame( cos2axis1 = mca$var$cos2[,1], cos2axis2 =
        mca$var$cos2[,2],cos2axis12 =(mca$var$cos2[,1]+mca$var$cos2[,2])),2)
```

```
qualitymca[rev(order(qualitymca$cos2axis12)), ]
```

	cos2axis1	cos2axis2	cos2axis12
1	0.70	0.27	0.97
Type_R	0.70	0.27	0.97
Type_D	0.70	0.27	0.97
3	0.00	0.90	0.90
Station_B	0.00	0.90	0.90
2	0.79	0.07	0.87
Station_A	0.79	0.07	0.87
I	0.26	0.53	0.79
E	0.26	0.53	0.79
Farmland	0.64	0.03	0.67
Station_D	0.13	0.28	0.41
Station_C	0.40	0.00	0.40
Urban	0.28	0.01	0.29
ter	0.02	0.20	0.22
sec	0.03	0.17	0.21
Grassland	0.10	0.01	0.11
prim	0.09	0.00	0.09
Yes	0.01	0.00	0.02
No	0.01	0.00	0.02

Ten of the 15 factors are well represented in the space formed by the first three axes of the MCA: the two types of marsh, the predominant land use as farmland in the neighboring area of influence, the three surfaces of the drainage basins, internal and external position and the four stations (>0.6).

4.1.4.3. Representation and interpretation of the results of the MCA

```
par(mfrow=c(2,2))
plot(mca,axes = c(1,2))
s.class(mca$ind$coord[,c(1,2)] , fac=fext$Station)
plot(mca,axes = c(1,3))
s.class(mca$ind$coord[,c(1,3)] , fac=fext$Station)
```

In unfed systems (negative values on axis 1), marsh A is characterized by a marked use of the land as farmland in the neighboring area of influence and by median drainage basin surfaces (Figure 4.12), whereas marsh B is characterized by large drainage basins (Figure 4.12). In re-fed systems (positive values on axis 1), marshes C and D are characterized by a predominant use of grassland or urban land in the neighboring area of influence and by small drainage basins (Figure 4.12).

A. Dual representation
(Axe 2 vs Axe 1)

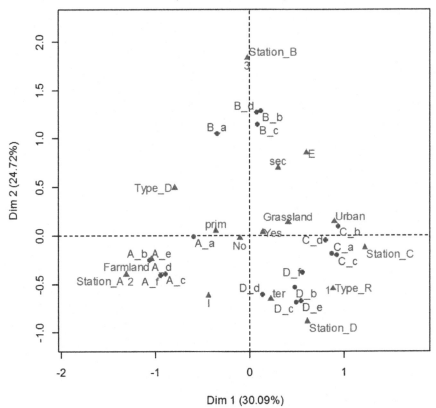

Figure 4.12. *Graphical representations of the MCA applied to the "environmental factors" database. Representation of the factors (•) and stations (▲) on the second axis versus the first axis. For a color version of this figure, see www.iste. co.uk/david/data.zip*

4.1.5. *Principal coordinates analysis*

4.1.5.1. *When should we carry out principal coordinates analysis?*

Principal coordinates analysis (PCoA) [TOR 58] represents a generalization of unconstrained ordination. In this sense, it allows us to carry out an **analysis with the distance or dissimilarity of our choice.** Therefore, the applicability conditions depend on the chosen distance/dissimilarity. If we use a symmetric coefficient, the applicability conditions and the interpretation are similar to those of a PCA (see section 4.1.2), whereas, if we employ an asymmetric coefficient, they are similar to those of a CA (see section 4.1.3). PCoA is always carried out in **direct mode**, just like any other kind of unconstrained analysis, and it requires **the number of objects to be greater than the number of parameters.**

PCoA can be carried out with the function *cmdscale()*, available as one of the core functionalities of the R programming language. We have to specify the association matrix based on the coefficient we have chosen and the number of axes we want to extract in the *k* argument. This analysis is applied to the log-transformed "phytoplankton" database processed with a Bray–Curtis dissimilarity matrix (MATphyto). The *eig=TRUE* argument provides access to the eigenvalues. We are dealing with an asymmetric coefficient, so the results will be interpreted as CA.

```
PCoA<-cmdscale(MATphyto,k=(nrow(phytoS)-1),eig=T)
PCoA #Different tables of available results
```

The help section for the *cmdscale()* function describes this as a multidimensional analysis, but it is not the case; the latter is presented in detail in section 4.2. The principle of this kind of analysis is based on a representation in a low-dimensional space (in general two dimensions) of the distances from an association matrix, rather than on the extraction of axes that summarize the information. Eigenvalues are extracted from this analysis, which proves how this is an unconstrained ordination, instead of a multidimensional analysis as it is used in [BOR 11].

4.1.5.2. *Quality indicators*

A) How many axes should we consider?

round(PCoA$eig,2)->eig ;eig *#Eigenvalues, some of them negative*

0.65 0.39 0.35 0.21 0.20 0.10 0.05 0.04 0.03 0.01 0.00 0.00 0.00 0.00 -0.01 -

0.02 -0.04 -0.05 -0.07

Some eigenvalues are negative. We should remove them when we want to calculate the quantity of inertia explained by each axis.

eig[eig>=0]->eigS ; eigS *#Remove negative eigenvalues*

round(eigS/sum(eigS)*100,0) *#Rounded-up explained inertia for each axis*

32 19 17 10 10 5 2 2 1 0 0 0 0 0

As was the case for CA, only two indices help us determine how many axes we should consider:

– a percentage of explained inertia equal to at least 50%. Here, the first two components already contain 51% of the information included in the database;

– a break in the gradient of the eigenvalues or of the explained inertia scree plot. Here, the change actually takes place between the third and the fourth axis (Figure 4.13).

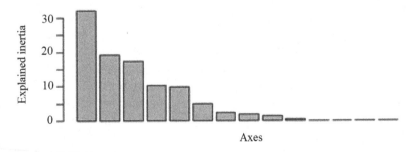

Figure 4.13. *Scree plot of the inertias explained by 14 axes of the PCoA carried out on the "phytoplankton" database*

Therefore, all the results seem to lead us to consider two or three axes. In simple terms, the exploitation will be made on the first two axes.

B) Which parameters can be interpreted?

Since the $\cos^2$ are no longer available, the best way of determining the interpretable variables is still to observe the parameters (i.e. taxa) with maximum coordinates on axes 1 and 2 with the *wascores()* function.

spe<-wascores(PCoA$points[,1:2],log1p(phytoS))

round(spe,2)

	V1	V2
Cyd	-0.06	0.02
Navi	0.01	0.01
Cyli	-0.13	0.07
Nitz	-0.05	-0.02
Gymn	0.01	0.09
Cosm	-0.16	0.03
Anki	0.01	-0.08
Mono	-0.14	0.02
Oocy	-0.11	0.02
Scen	-0.07	-0.03
Cryp	0.03	0.04
Eugl	-0.04	-0.03
Phac	-0.05	0.00
Stro	-0.08	-0.04
Trac	-0.04	-0.03

Only seven taxa out of 15 are well represented on the principal plane formed by the first two axes with coordinates whose absolute value is equal to or greater than 0.08:

– On axis 1: *Cosmarium, Monoraphidum, Oocystis and Strombidium*;

– On axis 2: *Gymnodinium, Ankistrodesmus*.

Four taxa – *Cosmarium, Monoraphidium, Oocystis* and *Strombidium* – influence the stations situated among the negative values of axis 1 (Figure 4.14(A)), *Gymnodinium* influences those situated among the positive values of axis 2, while *Ankistrodesmus* affects those situated among the negative value of axis 2 (Figure 4.14(A)).

4.1.5.3. Representation and interpretation of the PCoA

ordiplot(scores(PCoA)[,c(1,2)], type="t" ,xlab="Axis 1",ylab="Axis 2")

abline(h=0);abline(v=0)

text(spe,rownames(spe),cex=0.7,col="red")

s.class(PCoA$points[,c(1,2)] , fac=fext$Station,

s.class(PCoA$points[,c(1,2)] , fac=fext$Type

The non-re-fed stations are more influenced by the presence of the *Cosmarium, Monoraphidium, Oocystis* and *Strombidium* taxa, unlike the stations of re-fed marshes (Figure 4.14(C)). The stations of marsh C are influenced by the presence of *Gymnodinium*, and marsh A is affected by the presence of *Ankistrodesmus*, whereas the stations of marsh A are distributed distinctly on axis 2 and are consequently influenced by both taxa (Figure 4.14(B)).

Observation: The results are different from those of the CA applied to the same data matrix because of the distances/dissimilarities used: χ^2 distance for the CA and the Bray–Curtis dissimilarity for the PCoA. The latter is not a mere profile analysis, and it also takes into consideration the differences in abundances when creating the association matrix.

A. Dual representation
(Axis 2 vs Axis 1)

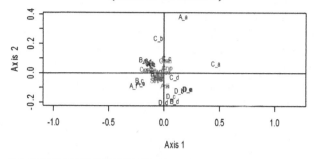

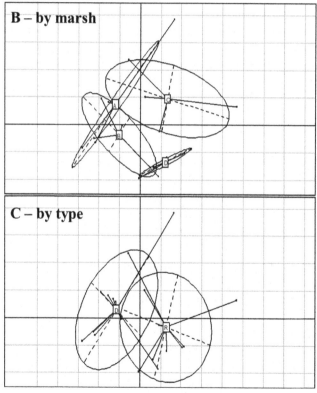

Figure 4.14. *Graphical representations of the PCoA applied to the simplified "phytoplankton" database. Representation of the taxa (grey) and stations (black) on the second axis versus the first axis (A). Dispersion of the stations per marsh (A–D) on the second axis versus the first axis (B) and dispersion of the stations per type of marsh (non-re-fed or re-fed) on the second axis versus the first one (C)*

4.2. Non-parametric alternative: nMDS

This type of analysis aims to **represent the distances between elements in a reduced space**, generally a 2D space [KRU 78]. nMDS is based on the ranks between these distances, rather than on the distances themselves. Therefore, this analysis does not require any **preliminary applicability condition**, and we refer to a non-parametric multivariate analysis. On the other hand, its interpretation depends on its ability to reduce the whole of the information contained in a matrix that includes x variables to two (or three) dimensions. This kind of analysis allows us to highlight groups and/or gradients. In this sense, it is more powerful than clustering analysis, which focuses on underscoring groups of objects. However, this type of analysis presents the same drawback as clustering analysis, since it cannot provide direct information on the variables that structure the elements, unlike unconstrained ordination. However, a quality index, i.e. stress, allows us to determine the representation quality. If this index is bad, the analysis cannot be interpreted.

4.2.1. Mathematical principle

The principle of multidimensional scaling is based on an iterative process. We have to choose the most intuitive number of dimensions, generally two axes, in order to be able to interpret the result on a plane (Figure 4.15). The elements are first randomly placed on this plane. Then, the Shepard diagram, which represents the distance between each pair of elements in relation to the distance in the association matrix, is built. Afterward, a regression line is adequately adjusted and the stress is calculated based on it (Figure 4.15). The position of the elements is slightly changed before recalculating the stress many times over (iterative process). The position kept is the one that yields the lowest stress. All of these operations, based on a random position of the elements at the beginning, are performed several times (by default, 20 times in R). The optimal position is obtained by choosing the operation that involves the least amount of stress (Figure 4.15). As its name indicates, nMDS is based on the ranks in the initial association and distance matrix on the plane of the final analysis.

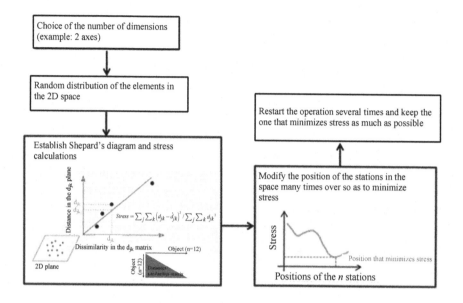

Figure 4.15. *Schematic representation of the implementation of multidimensional scaling*

A stress value of 0 represents a perfect adjustment. However, it may indicate that several solutions are possible, especially if the number of elements is low. An index value smaller than 0.05 gives an almost perfect representation of the elements in the space, a stress value between 0.05 and 0.1 provides a good representation, whereas a stress value between 0.1 and 0.2 must be considered more carefully, since certain elements run the risk of being badly represented. Finally, a stress value greater than 0.2 cannot be interpreted. Attention should be paid here, since no information is assigned to the axes, which was not the case for the factor analyses presented in section 4.1.

nMDS can be applied with the *metaMDS()* function, which belongs to the library [vegan]. It requires us to specify at least, in the k argument, the number of dimensions desired to represent the elements in the space (two or three). nMDS can be carried out directly based on an association matrix or the transformed database. In the latter case, it requires to specify, in the "distance" argument, the association coefficient we have chosen to build the association matrix. Finally, in order to avoid another transformation of the data, apart from the one carried out before the analysis, we have to add the "autotransform=FALSE" argument.

4.2.2. Direct mode

4.2.2.1. Based on an association matrix (e.g. phytoplankton communities)

The first nMDS performed will be carried out to determine the structure of the stations in the database of the phytoplankton communities (15 taxa). The association matrix, called "MATphyto", has been built beforehand with the log-transformed abundances. Since the nMDS is carried out directly on the "MATphyto" Bray–Curtis dissimilarity matrix, we do not have to specify any distance.

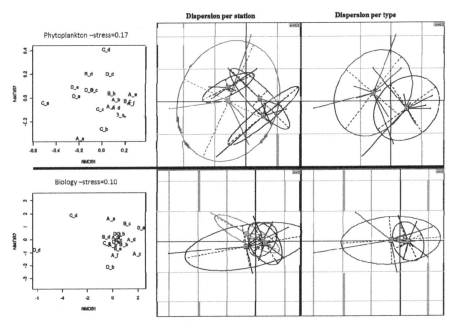

Figure 4.16. *Representation of the planes resulting from a 2D multidimensional scaling for the phytoplankton communities (above) and the biology database (below). The representation per station and marsh type has also been provided for each analysis*

library(vegan)

NMDSphyto<-metaMDS(MATphyto, k=2, autotransform=F)

NMDSphyto

NMDSphyto$stress

The stress value obtained here is equal to 0.17. Therefore, we have to pay attention to the local structures, even if the result can still be interpreted. Then, the *plot()* function represents the nMDS plane. The *s.class()* function, which belongs to the library [ade4], identifies graphically in a 2D space the dispersion of the objects in the form of a qualitative variable. Thus, this function explores the data graphically. For example, Figure 4.16 represents the dispersion of the elements by marsh (four marshes: A–D) and by type of marsh (D or R).

par(mfrow=c(2,3))

plot(NMDSphyto,type="n", main=paste ("Phytoplankton -stress",
        round(NMDSphyto$stress,2)))

Psites<-scores (NMDSphyto, display="sites")
        text(Psites,labels=rownames(Psites))

library(ade4); s.class(Psites, fac=fext$Station)

s.class(Psites, fac=fext$Type)

We can realize that the stations of re-fed marshes vary more than those of non-re-fed marshes in terms of phytoplankton communities (Figure 4.16). Besides, these communities seem to vary a little from one type of marsh to another, since the stations of non-re-fed marshes are situated farther to the left than the stations of the re-fed ones.

4.2.2.2. *Based on a database (for example biological parameters)*

The second nMDS performed is applied to the "Biology" database, which includes seven variables corresponding to diversity indices and production/

productivity values – they have been previously standardized, since variables were expressed in different units ("BIO.trf"). The *metaMDS()* function can also be applied to a transformed or untransformed database. In this case, it will be necessary to specify the association coefficient required to establish the association matrix on which the nMDS will proceed (argument "distance"). Here, we use the Euclidean distance, since it is the most suitable for the data set. However, any distance or dissimilarity available with the *vegdist()* function, belonging to the library [vegan], can be used.

#Performing nMDS

NMDSbio<-metaMDS(BIO.trf, k=2, distance="euclidean", autotransform=F)

NMDSbio$stress

#Graphic representations

plot(NMDSbio,type="n", main=paste ("Biology -stress", round(NMDSbio$stress,2)))

Bsites<-scores(NMDSbio,display="sites")
      text(Bsites,labels=rownames(Bsites))

s.class(Bsites, fac=fext$Station)

s.class(Bsites, fac=fext$Type)

The stress value is good (0.11) and, consequently, the nMDS graphical analysis can be interpreted. The stations of marshes A and B, which correspond to the non-re-fed marshes, are much closer to each other in terms of their diversity, production and productivity, unlike the re-fed stations, which are quite spread out (Figure 4.16).

4.2.3. Indirect mode

This type of analysis can also be carried out in indirect mode to highlight relationships between the variables (as was the case for clustering analysis).

Here, we use the "DISTphys" distance matrix to determine the relationships between the chemical–physical variables.

#Performing nMDS

NMDSphys<-metaMDS(DISTphys,k=2,autotransform=F)

NMDSphys$stress

#Graphic representations

par(mfrow=c(1,2))

plot(NMDSphys,main=paste ("Chemical physics -stress",
round(NMDSphys$stress,2)))
    Physvar<-scores(NMDSphys, display="sites")
    text(Physvar,labels=rownames(Physvar))

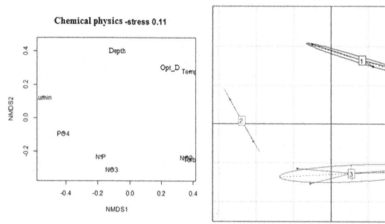

Figure 4.17. *Representation of the plane resulting from a 2D multidimensional scaling in indirect mode to establish the relationships between chemical–physical parameters. The representation per group of variables obtained with the UPGMA cluster analysis has also been provided*

Once again, the stress value is acceptable and the graphic analysis carried out on the basis of the nMDS plane is reliable enough (Figure 4.17). Here, the *s.class()* function allows us to observe how the groups obtained with the clustering analysis carried out on the same matrix lie on the nMDS plane. The variables that belong to the same group, because of the UPGMA clustering analysis that has been carried out, are in close proximity on the nMDS plane.

5

Understanding a Structure

This chapter aims to provide certain tools that allow us to link a structure like those highlighted in Chapter 3 or 4 (i.e. groups or gradients) to other factors (e.g. environmental factors). This relationship can be analyzed in structures observed in direct mode (i.e. structure of elements on the basis of chemical–physical (CP) parameters and biological parameters).

This relationship can be understood **without any causal hypothesis**, i.e. without *a priori* defining a specific set of variables as explanatory or explained (see section 5.1). This kind of analysis allows us, in particular, to investigate the similarities between *a priori* independent set of variables (e.g. climate vs. anthropogenic pressures). In this case, we can (1) compare how elements (e.g. stations/dates) are allocated to groups among several sets of variables (e.g. biological communities and CP variables through a χ^2 test, see section 5.1.1), (2) correlate different sets of variables by means of their association matrix (Mantel's test, see section 5.1.2) or the analysis of matrices of "raw" data with a parametric (large data sets, multiple factor analysis (MFA); see section 5.1.3.1) or non-parametric approach (small data sets, Procrustes analysis; see section 5.1.3.2).

However, most of the time, this kind of analysis is used **to explain a structure** that has been previously highlighted, namely to determine which factors generate this structure (e.g. CP parameters structuring biological communities, see section 5.2). We can consider how these structures account for the observation of groups by choosing the structuring variables (qualitative or quantitative) with decision trees (section 5.2.1.1) and quantifying their action with analyses of variance (ANOVAs, parametric or non-parametric, section 5.2.1.2) The quantitative factors that structure gradients may be analyzed with *a posteriori* correlations through permutation tests on the axes (see section 5.2.2.1.1) or by following the Bio-Env procedure (see section 5.2.2.1.2) when we are dealing with small data sets. When we are dealing with large data sets, these factors may be analyzed with active regressions through canonical analyses (section 5.2.2.2). Structuring qualitative factors can be studied with permutational ANOVA (PERMANOVA) analyses (see section 5.2.3.2) if we are dealing with small data sets, and with multivariate analysis of variance (MANOVA) and discriminant analyses (see section 5.2.3.1) if we are dealing with large data sets.

5.1. Correlating a structure with one or more structures without causal hypothesis

5.1.1. Correlating groups

Building a contingency table between the groups obtained with two classifications (e.g. the "chemical–physical" and "phytoplankton" databases), and applying a χ^2 test, allows us to determine whether the groups obtained are linked to one another [BRO 11]. The χ^2 H0 null hypothesis assumes that there is no relationship between the two groups obtained and, therefore, between the two databases, while the H1 alternative hypothesis assumes that there is a relationship.

The two classifications are first carried out with the UPGMA method: one of them on the Euclidean distance matrix obtained with the standardized "chemical–physical" database (MATphys), while the other on the Bray–Curtis dissimilarity matrix obtained with the simplified and log-transformed "phytoplankton" database (MATphyto).

```
par(mfrow=c(1,2))

cah.UPGMA.phys<-hclust(MATphys, method="average")

clust.phys<-reorder.hclust(cah.UPGMA.phys,MATphys)

plot(clust.phys, hang=-1, xlab="Groups of stations", ylab="Euclidean distances",
     main="Chemical physics")

rect.hclust(clust.phys,k=5, border="red")
```

phys.grpe5<- cutree(cah.UPGMA.phys, k=5); phys.grpe5 #*Each station is attributed to a group*

cah.UPGMA.phyto<-hclust(MATphyto, method="average") # UPGMA

clust.phyto<-reorder.hclust(cah.UPGMA.phyto,MATphyto)

plot(clust.phyto, hang=-1, xlab="Groups of stations", ylab="Bray-Curtis dissimilarities", main="Phytoplankton")

rect.hclust(clust.phyto,k=5, border="red")

phyto.grpe5<- cutree(cah.UPGMA.phyto, k=5);phyto.grpe5

The two classifications are optimized by using five groups. Allocating the stations among the five groups can thus be done between the two matrices with the χ^2 test on the contingency table.

table(as.factor(phys.grpe5),as.factor(phyto.grpe5))->table

plot(table)

chisq.test(as.factor(phys.grpe5),as.factor(phyto.grpe5))

Pearson's Chi-squared test

data: as.factor(phys.grpe5) and as.factor(phyto.grpe5)

X-squared = 33.0741, df = 16, *P*-value = 0.007225

The χ^2 test yields a P-value of 0.007, which allows us to reject the null hypothesis and to accept the alternative one, according to which there is a relationship between the two classifications, i.e. "chemical–physical" and "phytoplankton" databases. The representation of the contingency table can provide more information about this relationship (Figure 5.1). The groups obtained with the "chemical–physical" matrix are arranged into rows, whereas those obtained with the "phytoplankton" matrix are arranged into columns. Group 1 of the "chemical–physical" matrix (formed by stations A_a, A_c, B_b, B_c, B_d,C_a,C_b, C_c and D_d) is divided up among group 1 (A_a), group 2 (A_c, B_b, B_c, C_b, C_c), group 3 (B_d, D_d) and group 4 (C_a) of the "phytoplankton" matrix. Groups 2 and 3, obtained with the "phytoplankton" matrix, can be found in the same group. Group 4, which includes only one station, obtained with the "phytoplankton" matrix, corresponds perfectly to group 5, obtained with the "chemical–physical" matrix. Finally, group 5, obtained with the "phytoplankton" matrix, includes the stations of group 3 in the classification obtained with the "chemical–physical" matrix (D_a, D_b, D_c and D_e). Thus, the classifications obtained through the two matrices are not perfectly identical, even though the χ^2 test points out that they are significantly linked.

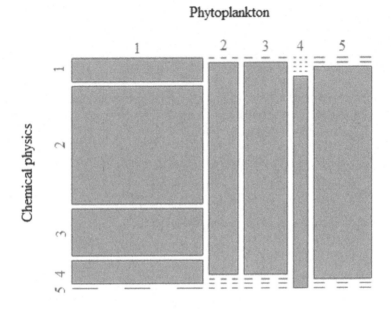

Figure 5.1. *Graphical view of the contingency table resulting from the comparison between the association matrices obtained based on the "chemical–physics" and "phytoplankton" databases*

5.1.2. Correlating association matrices

The Mantel test is a permutation test that allows us to study the linear correlation between two association matrices (distances, dissimilarities or similarities). The H0 null hypothesis assumes that the two matrices are not linked, while the H1 alternative hypothesis assumes that the matrices are linearly linked. The statistic test aims to determine a linear correlation coefficient between the two association matrices [MAN 67].

Here, we illustrate a Mantel test between the "environmental factors" and the different "chemical–physical", "biology" and "phytoplankton" matrices, as well as between the "chemical–physical" matrix and the "biology" and "phytoplankton" matrices, by applying Mantel tests on the matrices considered two by two. Let us first recall how these association matrices have been obtained.

```
library(ade4)

MATphys<-dist(CPtrf, method = "euclidean")

MATphyto<-vegdist(log1p(phytoS), method="bray")

MATbio<-dist(BIO.trf, method = "euclidean")

MATfext<-dist.binary(fext.disj[,c(6,8,9:19)], method=2)
```

As is the case for any permutation test, the results vary slightly each time the test is carried out. The principle of a permutation test (already developed for the SIMPROF test, see section 3.1.2.4) is to build a theoretical distribution curve by permuting the raw data a large number of times. These permutations create some "randomness" in the database. For each permutation, the statistic of the test is recalculated and the distribution curve of these values, created randomly, represents the curve with which the value of the statistic obtained for the database will be compared. If the statistic falls outside a 95% probability of obtaining the result that has been obtained randomly (95% of the surface of the theoretical distribution), the alternative hypothesis is accepted, otherwise the null hypothesis is kept. Therefore, the result depends on random permutations, which will vary each time we carry out the Mantel test. If the results vary too widely each time, we only have to add the *nrepet* argument to stabilize them. nrepet corresponds to the number of permutations carried out to obtain the theoretical frequency curve. This analysis may take some time, but the results will be more reliable.

```
mantel.randtest(MATfext,MATphys,nrepet=99999)->MT_FE_CP; MT_FE_CP
    #p-value=0,01*
mantel.randtest(MATfext,MATphyto,nrepet=99999)->MT_FE_PHY;
    MT_FE_PHY#p-value=0,009**
mantel.randtest(MATfext,MATbio,nrepet=99999)->MT_FE_BIO; MT_FE_BIO
    #p-value=0,57 ns
mantel.randtest(MATphys,MATphyto,nrepet=9999)->MT_CP_PHY;
    MT_CP_PHY#p-value=0,01*
mantel.randtest(MATphys,MATbio,nrepet=9999)->MT_CP_BIO; MT_CP_BIO
    #p-value=0,21 ns
```

```
x11();par(mfrow=c(3,2))
plot(MT_FE_CP,main="Environmental f.-Chemical physics")
plot(MT_FE_PHY,main="Environmental f.-Phytoplankton")
plot(MT_FE_BIO,main="Environmental f.-Biology")
plot(MT_CP_PHY,main="Chemical physics-Phytoplankton")
plot(MT_CP_BIO,main="Chemical physics-Biology")
```

Figure 5.2 shows, for each Mantel test carried out, a comparison between the statistic obtained with the raw database and the theoretical distribution curve obtained with 99,999 permutations. The results show that there are significant linear correlations between the "environmental factors" and the "chemical physics" matrices, as well as between these two matrices and the "phytoplankton" one. On the other hand, there seems to be no linear relationship between the "environmental factors", "chemical–physical" and "biology" matrices.

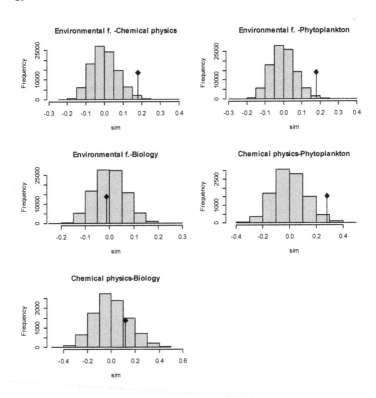

Figure 5.2. *Comparison between the statistic of Mantel's test obtained in the database with the theoretical distribution curve obtained for 99,999 permutations of the raw matrix for each test (matrices considered two by two)*

5.1.3. *Correlating different data tables*

5.1.3.1. *Parametric alternative: Multiple Factor Analysis (MFA)*

The MFA is a type of multivariate analysis that allows us to carry out a **correlative analysis between several data tables without any preassumption about the cause–effect relationships** between one table and another [ESC 94]: all tables are considered the same without taking into consideration any causal relationship. Several kinds of data tables may be used: tables of standardized (i.e. standard-score transformed) or unstandardized quantitative variables, tables of qualitative variables, etc. However, the kinds of variables within one table must be the same. This analysis is a parametric multivariate analysis governed by the same applicability conditions as factor analysis.

A principal component analysis (PCA; standardized or unstandardized) is carried out on each data table. Each PCA is then weighted, in order to give it the same amount of weight in the final analysis, by dividing each variable by the first eigenvalue of its PCA. Afterward, a global PCA is carried out on all the weighted PCAs.

To illustrate this type of analysis, we will consider three data tables: the table of "environmental factors" transformed into binary variables with a complete disjunctive table (see section 2.2.3), the table of standardized CP variables (see section 2.2.1) and the table of standardized biological variables (see section 3.1.3).

fextS<- fext.disj[,c(6,8,9:19)] names(fextS);dim(fextS)

CPtrf<-as.data.frame(scale(CP));names(CPtrf);dim(CPtrf)

BIO.trf<-as.data.frame(BIO.trf);names(BIO.trf);dim(BIO.trf)

The MFA is carried out here with the *MFA()* function, which belongs to the (FactoMineR) library. The tab database corresponds to the three concatenated matrices, the **group** argument allows us to dissociate the tab column number corresponding to each group of variables, the **ncp** argument the one corresponding to the number of axes considered (a maximum for now, i.e. a number corresponding to the number of variables, 18), while the **name.group** argument the one corresponding to the name of the groups of variables.

```
library(FactoMineR)
tab<-data.frame(fextS,CPtrf,BIO.trf); names(tab)
gr<-c(ncol(fextS),ncol(CPtrf),ncol(BIO.trf))
mfa<-MFA(tab, group=gr, type=c("c","s","s"), ncp=25, name.group=
    c("fextS","CPtrf","BIO.trf") ,graph=TRUE, axes = c(1,2))
mfa
```

There are several tables that can account for all the results. We only have to call the object in which these tables are stored (i.e. mfa), followed by $ and the name of the desired subset.

The quality indicators are the same as for unconstrained ordination.

```
head(round(mfa$eig,3))
```

	eigenvalue	percentage of variance	cumulative percentage of variance
comp 1	1.774	21.516	21.516
comp 2	1.543	18.716	40.232
comp 3	1.129	13.697	53.929
comp 4	1.024	12.424	66.354
comp 5	0.707	8.580	74.933
comp 6	0.507	6.153	81.087

Here, we use the same indicators used for PCA. Four axes present eigenvalues greater than 1. Three axes can explain on their own more than 50% of the total variance (54%), while the break in gradient takes place between the second and third axis. The compromise between the three indicators tells us to keep three axes.

Interpretable parameters

```
quality<-round(data.frame(cos2axis1=mfa$quanti.var$cos2[,1],
    cos2axis2=mfa$quanti.var$cos2[,2],cos2axis3=mfa$quanti.var$cos2[,3],co
    s2axis123=mfa$quanti.var$cos2[,1]+mfa$quanti.var$cos2[,2]+mfa$quanti.
    var$cos2[,3]),2)

quality[rev(order(quality$cos2axis123)),]
```

	cos2axis1	cos2axis2	cos2axis3	cos2axis123
NO3	0.34	0.48	0.00	0.82
Depth	0.30	0.02	0.46	0.78
DB.1	0.68	0.08	0.00	0.76
Type.R	0.68	0.08	0.00	0.76
Opt_Depth	0.46	0.20	0.04	0.71
NO2	0.06	0.55	0.07	0.68
Temp	0.49	0.18	0.01	0.68
Channel.prim	0.07	0.06	0.56	0.68
P.B	0.09	0.00	0.57	0.66
N.P	0.14	0.50	0.00	0.65
Land_use.Urban	0.10	0.04	0.52	0.65
Land_use.Grassland	0.16	0.04	0.44	0.63
Land_use.Farmland	0.46	0.13	0.00	0.59
Position.I	0.03	0.51	0.02	0.56
RS	0.19	0.35	0.00	0.54
D_Simpson	0.32	0.18	0.03	0.54
Turb	0.01	0.42	0.12	0.54
DB.2	0.53	0.00	0.00	0.54
H_Shannon	0.38	0.12	0.02	0.52
P_phyto	0.04	0.22	0.17	0.43
PO4	0.01	0.29	0.13	0.43
MP.Yes	0.12	0.02	0.18	0.33
MP.No	0.12	0.02	0.18	0.33
Channel.ter	0.04	0.03	0.25	0.33
Lumin	0.01	0.12	0.19	0.32

Showing 1 to 25 of 29 entries

The parameters that can be interpreted present cos^2 values (> 0.60 on the first three cumulative axes (arbitrary threshold)).

library(ade4)

x11();par(mfrow=c(2,2))

plot(mfa,choix="var",axes = c(1,2),cex=0.5)

plot(mfa,choix="var",axes = c(1,3),cex=0.5)

s.class(mfaindcoord[,c(1,2)] , fac=fext$Station)

s.class(mfaindcoord[,c(1,3)] , fac=fext$Station)

On axis 1, small drainage basins, re-fed marshes and optical depth (positive values) lie opposite average-sized drainage basins, the predominant use of land as farmland in the neighboring area of influence, water temperature and the Shannon diversity (negative values; Figure 5.3(A)). Axis 2 shows, among the positive values, internal stations, concentrations of nitrites, the N/P ratio and turbidity (Figure 5.3(A)). On axis 3, primary channels, the predominant use of land as urban land in the neighboring area of influence (positive values) lie opposite the predominant use of land as grassland in the neighboring area of influence and productivity (negative values; Figure 5.3(B)). The concentrations of nitrates present medium positive values between axes 1 and 2, while depth presents positive medium values between axes 1 and 3.

Marsh A and marsh B (negative values) lie opposite marsh C and marsh D (positive values) on axis 1 (Figures 5.3(C) and (E)). Marsh D presents positive values on axis 2, unlike marshes B and C (Figures 5.3(C) and (E)). The stations of the different marshes are distributed along axis 3 (Figures 5.3(D) and (F)). Therefore, the stations of marsh D, re-fed by the Charente, present small-sized drainage basins and are characterized by high concentrations of nitrates, high N/P ratios, high turbidities and low Shannon diversity of phytoplankton. The stations of marsh C, re-fed by the Charente, are situated in an external position and characterized by high optical depth and water temperature, and low nitrites and turbidities. The stations of marsh B, unfed by the Charente, present low concentrations of nitrates, a low N/P ratio, low turbidity, medium- or large-sized drainage basins and high Shannon diversity values of phytoplankton. Finally, the stations of marsh A, unfed by the Charente, present medium-sized drainage basins, a predominant use of land as farmland in the neighboring area of influence, low optical depth and high Shannon diversity of phytoplankton. For each type of marsh, the stations are distributed along this gradient: "high phytoplankton productivity linked to a predominant use of land as grassland in the neighboring area of influence versus predominant use of land as urban land with low depth".

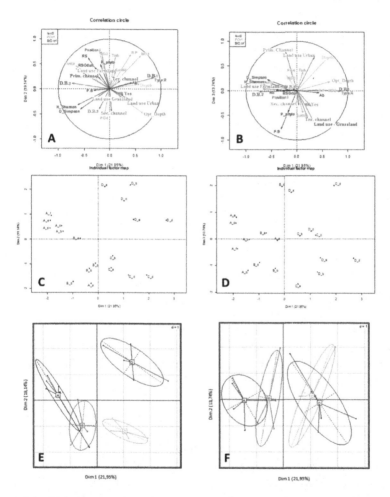

Figure 5.3. *Global correlation circle of the multiple correspondence analysis of the second axis versus the first axis (A) and the third axis versus the first axis (B). Representation of the stations of the second axis versus the first axis (C) and the third axis versus the first axis (D). Dispersion of the stations per marsh of the second axis versus the first axis (E) and the third axis versus the first axis (F). For a color version of this figure, see www.iste.co.uk/david/data.zip*

```
rvp<-mfa$group$RV
rvp<-rvp[-4,-4]
rvp[1,2]<-coeffRV(scale(fextS),scale(CPtrf))$p.value
rvp[1,3]<-coeffRV(scale(fextS),scale(BIO.trf))$p.value
rvp[2,3]<-coeffRV(scale(CPtrf),scale(BIO.trf))$p.value
round(rvp,3)
```

	fextS ≑	CPtrf ≑	BIO.trf ≑
fextS	1.000	0.007	0.495
CPtrf	0.437	1.000	0.019
BIO.trf	0.221	0.352	1.000

The **rpv** object accounts for the correlations between the sets of variables. The triangle in the bottom left-hand corner provides the correlation coefficients, while the triangle in the top right-hand corner provides the *P*-values. In this case, it seems that the environmental factors database is significantly correlated with the chemical–physical database, but not to the biological matrix, whereas the chemical–physical database is significantly correlated with the biological database.

5.1.3.2. *Non-parametric alternative: generalized Procrustes analysis*

Generalized Procrustes analysis allows us to carry out a **correlative analysis between several data tables without any presuppositions about the cause–effect relationships** between one table and another and without preliminary applicability conditions [GOW 75].

This analysis represents a technique used to compare forms. It consists of rotating a configuration (e.g. a multidimensional scaling analyses) to a maximum similarity with another comparable configuration choosen as reference –the target matrix. A test is then conducted to analyse the non-randomness ('significance') between the two configurations. As for nMDS, this type of analysis can be applied to small data sets. Procrustes analysis is carried out in R with the ***Procrustes()*** function while the associated test is conducted with the ***protest()*** function, both available in the [vegan] library.

We illustrate the procrustes analysis by comparing the association matrices of the biology and the Chemical-Physics databases to the Phytoplankton one (MATbio, MATphys and Matphyto, respectively). The

Understanding a Structure   107

latter is used as the target matrix. Note that the association matrices had been built using different association coefficients (Euclidean distance for the Biology and the Chemical-Physics databases *versus* Bray-Curtis dissimilarities for the phytoplankton database). nMDS were first realized for each matrix as follows:

nmds_phy<-metaMDS(MATphyto, k=2, autotransform=F) ;
nmds_phy$stress
[1] 0.1674129
nmds_bio<-metaMDS(MATbio, k=2, autotransform=F) ; nmds_bio$stress
[1] 0.1146576
nmds_phys<-metaMDS(MATphys, k=2, autotransform=F) ;
nmds_phys$stress

[1] 0.146768

x11();par(mfrow=c(3,1))
plot(scores(nmds_phy), main=paste("Phytoplankton - stress=",
round(nmds_phy$stress,3)),type="n");
        text(scores(nmds_phy),labels=rownames(scores(nmds_phy)))
plot(scores(nmds_bio), main=paste("Biology - stress=",
round(nmds_bio$stress,3)),type="n");
        text(scores(nmds_bio),labels=rownames(scores(nmds_bio)))
plot(scores(nmds_phys), main=paste("Chemical physics - stress=",
round(nmds_phys$stress,3)),type="n");
        text(scores(nmds_phys),labels=rownames(scores(nmds_phys)))

The stresses were acceptable for the three nMDS (Figure 5.4.), the procruste analysis could then be conducted.

library(vegan)
vare_phybio <- procrustes(X=nmds_phy,Y=nmds_bio) #*Procruste analysis of the biology database based on the phytoplankton database*
test_phybio<-protest(X=nmds_phy,Y=nmds_bio, scores = "sites", permutations = how(nperm = 99999))
test_phybio #Procruste tests
Procrustes Sum of Squares (m12 squared):   0.9436
Correlation in a symmetric Procrustes rotation: 0.2375
Significance: 0.58785

Permutation: free
Number of permutations: 99999

vare_phyphys <- procrustes(X=nmds_phy,Y=nmds_phys); *#Procruste analysis of the Chemical-physics database based on the phytoplankton database* testphyphys<-protest(X=nmds_phy,Y=nmds_phys, scores = "sites", permutations = how(nperm = 99999))
testphyphys
Procrustes Sum of Squares (m12 squared):   0.8673
Correlation in a symmetric Procrustes rotation: 0.3642
Significance: 0.17573

Permutation: free
Number of permutations: 99999

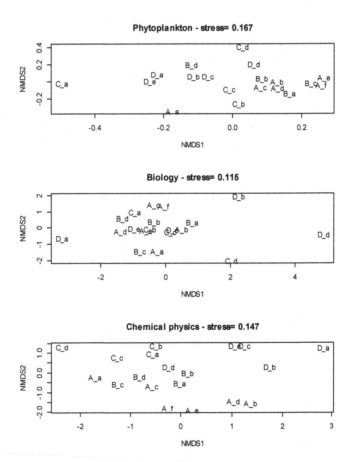

Figure 5.4. nMDS conducted over the three databases: Phytoplankton, Biology and Chemical-Physics

None of the rotated nMDS are significant since p-value is < 0.05. The rotated configurations can be represented as follows (Figure 5.5):

Rnmdsphybio<-as.data.frame(vare_phybio$Yrot) #*Scores of the rotated nMDS for Biology database*
Rnmdsphyphys<-as.data.frame(vare_phyphys$Yrot) #*Scores of the rotated nMDS for Chemical-Physics database*

x11();par(mfrow=c(2,1))
plot(Rnmdsphybio,type="n", main=paste("rotated Biology -Procrust test
    p=",round(test_phybio$signif,4)));text(Rnmdsphybio,labels=rowna
    mes(Rnmdsphybio),cex=0.8)
plot(Rnmdsphyphys,type="n", main=paste("rotated CP -Procrust test
    p=",round(testphyphys$signif,4)));text(Rnmdsphyphys,labels=rowna
    mes(Rnmdsphyphys),cex=0.8)

rotated Biology -Procrust test p= 0.5878

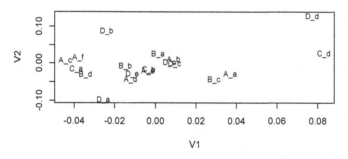

rotated CP -Procrust test p= 0.1757

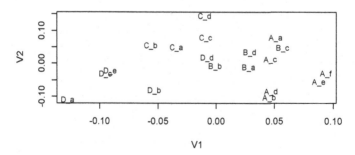

Figure 5.5. *Visualisation of the rotated nMDS for Biology and Chemical-physics databases*

5.2. Explaining a structure

5.2.1. The structuring factors of groups

5.2.1.1. Decision trees

A decision tree is a tool that **represents a set of choices in the shape of a tree**: the different possible "decisions" are situated at the end of the branches (the "leaves" of the tree) and are reached in relation to the decisions taken at each stage. There are several algorithms that can be used to build these trees (e.g. Classification and Regression Trees [CART]). They represent a supervised learning technique: a first set of data is used to build the tree, and a second set allows us to validate it (i.e. estimating the re-substitution success rate obtained when using it). The tree can then be used to extrapolate the value of the explained variable to a different set of data (i.e. prediction). One of the input variables is chosen at each node. We will distinguish between **classification trees**, predicting which of the different modalities (or classes) a qualitative variable belongs to, and **regression trees**, which predict the values of a quantitative variable.

The decision tree presented here correspond to the use of the **CART algorithm** [BRE 84]. This decision tree is **binary** in the sense that each node can only have two threads, and the segmentation criteria is the **Gini index**. This index is the statistical measure of the dispersion of elements in a given population, which corresponds to the relative mean absolute difference (mean absolute difference divided by the mean). Thus, it varies between 0 and 1, the latter value indicates that the elements of the population are perfectly equal. The goal of this method is to find the variable and a threshold related to each node that minimizes intraclass inertia or maximizes interclass inertia. Its effectiveness depends on the size of the learning database: the higher the number of elements contained in the data set are, the more reliable the tree is. Concerning classification trees, the number of potentially explanatory qualitative variables with many classes should be limited, since they tend to be chosen the most, given that they create a phenomenon of overfitting. Therefore, they should not be considered, or their number of classes should be reduced.

Here, we will only illustrate classification trees, since our objective is to explain how groups are formed (qualitative variables that are *a priori* or highlighted by cluster analysis) by quantitative or qualitative variables. The decision trees that employ the CART algorithm (classification or regression) can be built with the *rpart()* function, which belongs to the [rpart] library.

We will illustrate how it is used for the prediction of the groups of stations obtained with the SIMPROF test on the database of the biological communities based on chemical–physical variables and environmental factors (i.e. classification tree based on both quantitative and qualitative variables). The "group" variable and the potentially explanatory variables are first concatenated in the same "dat" data set. The "Marsh" and "DB" qualitative variables are not considered, since their number of classes is high.

library(rpart)

data.frame(SP=as.factor(simprof.phyto),CP,fext[,-c(1,6,8)])->dat

The elements (i.e. stations) are split in two: a first "app" data set, which allows us to build the tree (e.g. all the stations apart from 1, 2, 7, 10 and 15), and a second "val" data set, which enables us to validate it (stations 1, 2, 7, 10 and 15). These stations have been chosen randomly. Then, the tree is built with the "app" data set by specifying the qualitative variable that has to be explained; ~ the potentially explanatory variables (e.g. all by ~); the database in the *data* argument (e.g. all except for the variable explained in "app") and the method in the *method* argument (e.g. "class" for a classification tree).

names(dat)

app<-dat[-c(1,2,7,10,15),]

val<-dat[c(1,2,7,10,15),]

model = rpart(app[,1] ~ ., data = app[,-1], method = "class"); model

n= 14
node), split, n, loss, yval, (yprob)
* denotes terminal node
1) root 14 7 5 (0.071 0 0.36 0.071 0.5) *

We considered 14 elements when building the tree ($n = 14$). The asterisk (or asterisks) indicates the terminal node (or nodes). Here, there is only one of them. This means that no tree has been built. In this case, we have used the default parameters of the *rpart()* function, which does not build the most complete tree on the grounds of effectiveness. More specifically, it does not build trees from a data set including 20 or fewer elements, and it requires a relative improvement of at least 1% of the quality of a partition to perform a split. Besides, a very deep tree, which makes no mistakes in the classification, is not suitable due to its substantial overfitting and will have to be pruned. When we are dealing with small data sets, we can modify the base parameters by adding the *rpart.control()* function to the *control* argument. This argument requires us to specify the minimum number of elements (for example 5) below which it can continue splitting ("minsplit" parameter) and the constraint on the splitting quality ("cp" argument; for example 0%).

model2   <-   rpart(app[,1]   ~   .,   data   =   app[,-1],   method   = "class",control=rpart.control(minsplit = 3, cp = 0))

model2

n= 14
node), split, n, loss, yval, (yprob)
* denotes terminal node
1) root 14 7 5 (0.071 0 0.36 0.071 0.5)
      2) Type=R 8 3 3 (0.12 0 0.62 0.12 0.12)
            4) Land use= Farmland, Grassland 5 0 3 (0 0 1 0 0) *
            5) Land use= Urban 3 2 1 (0.33 0 0 0.33 0.33)
                  10) Depth>=0.885 2 1 1 (0.5 0 0 0 0.5) *
                  11) Depth< 0.885 1 0 4 (0 0 0 1 0) *
      3) Type=D 6 0 5 (0 0 0 0 1) *

The tree built includes four leaves (four asterisks). For each branch, we obtain three values which correspond, in this order, to the number of elements included in the branch (e.g. six for branch 3), the number of elements that have been badly classified (e.g. zero for branch 3) and the class of the qualitative variable explained (e.g. class 5 for branch 3). Each branch is defined by a quantitative variable threshold or a qualitative variable modality (for example Depth >=0.855 for branch 10, or class "re-fed type" for branch 2).

printcp(model2)
Classification tree:
rpart(formula = app[, 1] ~ ., data = app[, -1], method = "class",
    control = rpart.control(minsplit = 3, cp = 0))

Variables actually used in tree construction:
[1] Depth   Land_use Type

Root node error: 7/14 = 0.5

n= 14

```
      CP nsplit rel error  xerror   xstd
1 0.57143     0  1.00000 1.28571 0.25612
2 0.14286     1  0.42857 0.57143 0.24147
3 0.00000     3  0.14286 0.71429 0.25612
```

The prediction error at the root is equal to 50% (7/14) and three variables are employed for the tree: two qualitative variables – land use and marsh type – and one quantitative variable, i.e. depth.

As the number of leaves increases, the performance of the model improves at first and then decreases because of overfitting. Therefore, we have to choose the complexity that minimizes the estimated error (xerror), i.e. in this case cp = 0.14286. To automate the process by which we obtain this value,

OptimalModel<- prune(model2,
        cp=model2$cptable[which.min(model2$cptable[,4]),1])
OptimalModel

n= 14

node), split, n, loss, yval, (yprob)
* denotes terminal node

1) root 14 7 5 (0.071 0 0.36 0.071 0.5)
    2) Type=R 8 3 3 (0.12 0 0.62 0.12 0.12) *
    3) Type=D 6 0 5 (0 0 0 0 1) *

library(rpart.plot)

prp(OptimalModel,extra=1)

In the graphical representation of the pruned tree (Figure 5.6), each node corresponds to one question (for example type "refed"?), the true answer is on the left, while the false answer is on the right. For each branch, this model lists the main attribution class (group 3 on the left and group 5 on the right) as well as the number of attributions in this class through the learning tree of the elements resulting from different classes (e.g. if we consider the left branch, one element resulting from group 1, zero elements from group 2, five elements from group 3 and one element from groups 4 and 5).

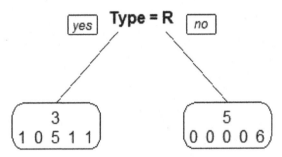

Figure 5.6. *Representation of the pruned classification tree explaining the preestablished groups of stations on the basis of the phytoplankton communities (with agglomerative hierarchical clustering). Only the "type of marsh" qualitative variable seems involved in the discrimination of groups 3 and 5*

The tree can then be validated with another data set (e.g. val), and the *predict()* function can assess the classes of the group variable in relation to the tree. A contingency table allows us then to compare the results obtained with the tree and the results known for each element. The re-substitution success rate is then calculated.

prev = predict(OptimalModel, newdata = val[,-1], type = "class")

mat = table(as.factor(prev), val[,1])

rate = sum(diag(mat))/sum(mat);rate

[1] 0.6

In this case, the re-substitution success rate is good (60%). However, this rate depends on the class considered. We obtain a 100% re-substitution rate for class 5 and a much lower one for the others.

prev = predict(OptimalModel, newdata = val[,-1], type = "class")

mat = table(as.factor(prev), val[,1])

Finally, if this rate is good, the tree can be used to make predictions with the *predict()* function.

This analysis may be used to simply select some structuring variables out of a set of potentially explanatory variables. For example, in this case the type of marsh represents the variable that best differentiates the groups determined on the basis of the phytoplankton communities, out of the set of the several qualitative and quantitative environmental variables considered.

5.2.1.2. Analysis of variance

ANOVAs allow us to explain a quantitative variable with one or more qualitative variables. In this sense, they allow us to **analyze the environmental variables that present significant differences among the groups** highlighted by cluster analysis, e.g. the groups of stations that present different phytoplankton communities.

These analyses are all based on a null hypothesis, according to which sampling fluctuation accounts for the differences between the groups, i.e. the groups present no significant differences on a population scale, and an alternative hypothesis, according to which sampling fluctuation cannot explain everything, i.e. at least one of the groups presents a significant difference on a population scale. Hence, rejecting the null hypothesis or accepting the alternative one involves a test, carried out afterwards to highlight the differences between different groups (e.g. post hoc for classical ANOVAs).

There are **several alternative types of ANOVAs**. They present different **power** and **robustness** (see section I.2.2 for these two notions) in relation to the mathematical principles of the tests used:

– classical ANOVAs are linear models with drastic applicability conditions, due to the use of the least squares method. Therefore, they are weak in terms of robustness but powerful;

– the ANOVAs based on generalized linear models are a little more robust;

– permutation ANOVAs, even more robust but less powerful;

– non-parametric ANOVAs subject to no applicability condition but much less powerful.

A specific alternative is chosen in relation to how powerful it is. It is advisable to use the most powerful test, i.e. the test that is most likely to lead us to reject H0. On the other hand, if the applicability conditions are not respected, a more robust alternative will be chosen.

Here, these analyses will be illustrated in relation to the groups obtained with a SIMPROF analysis carried out on the phytoplankton communities (*simprof.phyto* vector). As an example, the ANOVA's will be carried out for a chemical–physical parameter, i.e. the concentration of nitrates (**CP\$NO3**; Figure 5.7).

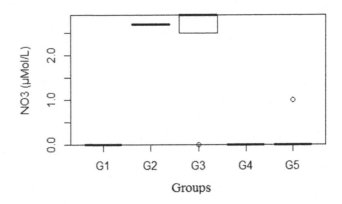

Figure 5.7. *Concentrations of nitrates according to the different groups of stations highlighted by the SIMPROF test on the phytoplankton communities*

boxplot(CP\$NO3~as.factor(simprof.phyto), xlab="Groups",ylab="NO3 (µMol/L)")

5.2.1.2.1. Parametric alternative: linear models

Linear models assume that any relationship between the explained and the explanatory variable has a linear shape. The applicability conditions are applied to the residuals of the models (i.e. differences, for each element, between the raw value and the result given by the model), rather than to the raw data [AZA 13]:

– **the residuals of the model are standardized**, i.e. homogeneously shared on both sides of the zero along the whole variation range of the variable explained;

– **the residuals must follow the normal distribution**. However, linear models are robust when we consider more than 20–30 elements;

– **the variances of the residuals are homogeneous** along the whole variation range of the variable explained. This is commonly the hardest condition to respect;

– **the residuals are independent**. This hypothesis responds to a simple random form of sampling. Any type of spatial (with a non-random position of the stations) or temporal sampling (with a uniform-distance form of sampling) may create a spatial or temporal autocorrelation. If this hypothesis is not respected, the "space" or "time" variable may be introduced into the model as a variable including a pairing (as a random variable).

Here, we will discuss classical linear models with the least squares method.

The model is adjusted so that the sum of squared residuals between each raw value and the value given by the model for all the values of the variable explained is minimal (i.e. method of "least squares"; [AZA 12]). According to this method, the sum of squared residuals from the general mean of the Y quantitative variable to be explained is equal to the sum of squared residuals between the modalities of the X explanatory qualitative variable and the intramodality one (modality A + modality B, etc.; Figure 5.8). The statistic of the test corresponds to a Fisher statistic, which compares inter- and intramodality mean squares. The value obtained must be compared with the one yielded by the Fisher–Snedecor known theoretical probability distribution, which represents the distribution related to the null hypothesis. If the F falls outside 95% of the values of this distribution, then H0 is rejected and H1 is accepted, i.e. at least one of the modalities differs from the others (Figure 5.8). The coefficient of determination (intermodality sum of the squares/total sum of the squares) yields the quantity of variance of the quantitative variable explained by the qualitative variable.

The linear model can be obtained with the *lm()* function by transforming the group variable into factor (*as.factor()* function).

lm(CP$NO3~as.factor(simprof.phyto))->AN

The applicability conditions are verified graphically in the graphs of the residuals in relation to the groups and the normal probability curve. The graphs required for the analysis of the applicability conditions can be obtained using the *plot()* function on the model.

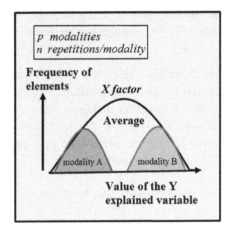

H0: The Y values are the same across modalitites

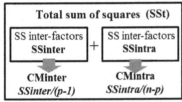

F Statistic = CMinter/CMintra

To be compated with the F of Fishner-Snedecor's law

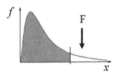

H1 accepted: at least one of the modalitites differs from the others

Figure 5.8. *Statistical development of an analysis of variance based on the least squares method. Two samples of the Y explained variable corresponding to the A and B modalities of an X qualitative factor are compared*

par(mfrow=c(2,2)); plot(AN) # *The graph window is split into 4 parts in order to display the 4 graphs yielded by the program*

The graph of the residuals in relation to the predicted values allows us to verify (1) whether the residuals are standardized – residuals substantially distributed around 0 – (2) the independence of the residuals (no banana shapes) and (3) the homogeneity of the variances of the residuals (no trumpet shapes on the right or on the left, Figure 5.9). The normal probability curve allows us to verify the normality of the residuals: the values should not get further from the bisector of the standardized residuals in relation to the quantiles (Figure 5.9). In the example considered here, the normality of the residuals does not seem to be respected (Figure 5.10(B)).

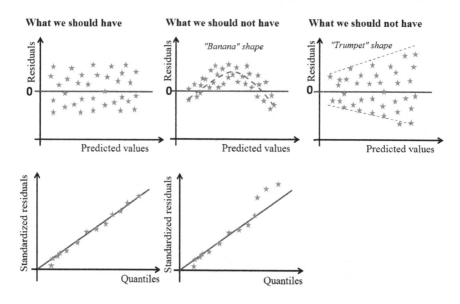

Figure 5.9. *Residuals' graph in relation to the predicted values and the normal probability curve. What we should have versus what we should not have*

The Shapiro–Wilk test enables us to verify more objectively the normality of the residuals, which are first retrieved from the model. The null hypothesis assumes that the residuals follow the normal distribution, whereas the alternative hypothesis assumes that this is not the case.

AN$res *#Residuals of the model*
shapiro.test(AN$res) *#Shapiro-Wilk test on the residuals*

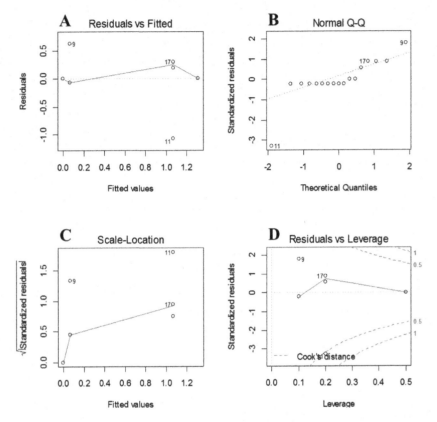

Figure 5.10. *Graphs allowing us to analyze graphically the applicability conditions of the ANOVA linear model. Only the two graphs (A) and (B) must be analyzed: the residuals in relation to the predicted values (A) and the normal distribution curve (B)*

In our example, the *P*-value is equal to 0.0001. H0 is therefore rejected and the normality of the residuals is not respected.

Levene's test allows us to test the homogeneity of the variances with the *leveneTest()* function, which belongs to the [car] library. The null hypothesis assumes that the variances of the residuals are the same, whereas the alternative hypothesis assumes that this is not the case.

```
library(car)
```

```
leveneTest(AN$res, as.factor(simprof.phyto)) #Levene's test on the residuals
```

In the example considered, the P-value yields 0.65. H0 is therefore kept and the homogeneity of the variances of the residuals is respected, as we can see in Figure 5.10(A).

The most effective way to test the normality and homogeneity of the variances of the residuals is graphically (Figure 5.10), since the Shapiro–Wilk test and Levene's test are less effective on the residuals than on raw data [AZA 12].

Thus, the applicability conditions are not respected in our example. If this had been the case, the P-value and the coefficient of determination of the model could have been retrieved with the *summary()* function. The coefficient of determination represents the quantity of variance explained by the linear model (R^2). If we were dealing with a P-value of less than 0.05, which would lead us to conclude that at least one group is significantly different from the others in terms of concentrations of nitrates, we could have carried out a multiple comparisons test with the *pairwise.t.test()* function to determine which groups were different.

summary(AN)

pairwise.t.test(CP$NO3, as.factor(simprof.phyto), p.adjust="bonferroni", pool.sd = T)

tapply(CP$NO3, INDEX= as.factor(simprof.phyto), FUN=mean) *#It allows us to retrieve the means by group*

When the number of qualitative factors increases, equations become more complex. These equations take into consideration the iteration number by type of the different factors, whether the factors are fixed (i.e. established by a researcher as part of his sampling strategy) and/or random (i.e. drawn by lot out of the set of possible factors, taking random values due to the sampling itself), whether the factors are crossed (i.e. each of them has a sense independently of the others) or hierarchized (i.e. the modality of the hierarchized factor B in factor A means nothing concrete as long as we do not know the modality associated with factor A), and whether the interaction between factors is considered or not (i.e. whether factor A evolves together with factor B or not). The implementation in R is trickier, since the help files sometimes provide the wrong information: consulting Azaïs and Bardet's manual [AZA 12] is recommended if we want to use it correctly. Besides, BIC models allow us to optimize the number of factors and interactions between significant factors [AZA 12], i.e. compromising between the number and the factor used and the coefficient of determination of the final linear model (namely, its explanatory power).

5.2.1.2.2. Intermediate alternative: PERMANOVA

PERMANOVA allows us to test the **response of one (or more) variable(s) to one or more qualitative factors** [AND 01]. Here, we only present the univariate case (see section 5.2.3.2 for the multivariate case). This analysis, based on an association matrix, is more robust than classical ANOVA in relation to the normality and homogeneity of the variances of the residuals [CHA 08]. However, it is sensitive on a non-equirepeated level (i.e. the number of elements by factors is different [AND 13]).

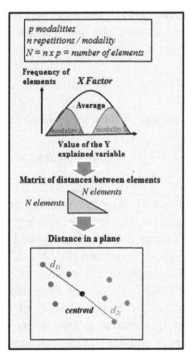

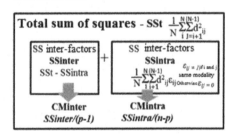

Figure 5.11. *Statistical development of a permutational analysis of variance. Two samples of the Y explained variable corresponding to the A and B modalities of an X qualitative factor are compared*

The principle of PERMANOVA is quite similar to the one of least-squares ANOVA based on the fact that the sum of squares can be calculated by using the distances between points in an n-dimensional space [AND 01]. For the case considered here, where one quantitative variable is to be explained, the distance matrix (i.e. the Euclidean distance) is calculated according to the same principle followed for several variables and in relation to the type of variable and the objectives (see section 2.3.1). The distance matrix lists the similarities/dissimilarities between elements (for example stations; Figure 5.11). For a Euclidean distance, the total sum of squares is calculated in the same way as the sum of the distances to the centroid (i.e. mean of the distances within the variables), the intramodality or residual sum of squares represents the sum for the set of modalities of the sums of squares to the centroid, while the intermodality sum of squares is the difference between the two previous sums ([MCA 01]; Figure 5.11). The mean squares can be obtained by dividing by the degree of freedom (modality number – 1 for intermodality MQ and number of elements – group number for intramodality MQ). As was the case for least-squares ANOVA, the statistic calculated is a pseudo-F corresponding to the intermodality mean square over the intramodality mean square. This statistic is compared with a theoretical distribution of the pseudo-Fs in relation to H0 calculated by permuting the initial database many times over. If the pseudo-F falls outside 95% of this theoretical distribution, H0 is rejected and H1 is accepted, i.e. at least one of the modalities of the factor differs from the others. The PERMANOVA provides then a pseudo-coefficient of determination (intermodality sum of squares/total sum of squares), yielding the quantity of the quantitative variable explained by the distance between groups of the qualitative factor.

For different distances, the distance to the centroid is harder to obtain. Thus, other methods can be used.

In our example, the analysis is carried out with a Euclidean distance matrix based on the concentrations of nitrates in 19 stations.

```
dist<- dist(CP$NO3, method= « euclidean »)
```

The analysis needs to test the homogeneity of the multivariate dispersions among groups. The betadisper test is performed with the *betadisper()* function, which belongs to the [vegan] library.

betadisper(dist(CP$NO3), as.factor(simprof.phyto), type = "centroid")->CA

anova(CA) # *It yields the result of the test*

In this example, the *P*-value yields 0.08 and implies that we should keep H0. The homogeneity of the dispersion among groups is therefore respected. The PERMANOVA can thus be carried out.

The PERMANOVA is implemented in R in the [vegan] library with the *adonis()* function. Here, it is carried out on a simple one-factor plane – the groups of phytoplankton communities – for a quantitative variable, namely the concentration of nitrates. It requires us to specify the quantitative variable ~, the association measure chosen in the argument *method* and finally the permutation number used to calculate the theoretical distribution in the *perm* argument. All the coefficients available in the *vegdist()* function belonging to the [vegan] library can be used. The Euclidean distance is used as an association measure for the CP variables.

adonis(CP$NO3   ~   as.factor(simprof.phyto),   method   =   "euclidean", permutations=999)->PERMANOVA

PERMANOVA #*Results are presented as an ANOVA table*

hist(PERMANOVA$f.perms) # *To obtain the distribution of the pseudo-f's based on H0*

The *P*-value is equal to 0.003 and H0 is rejected, i.e. at least one of the groups presents a concentration of nitrates that differs from the others. The groups explain 74% (pseudo-R^2) of the Euclidean distances among the groups. It is possible to test the differences between groups with the *TukeyHSD()* function.

TukeyHSD(CA)$group

	diff	lwr	upr	p adj
2-1	-2.220446e-16	-1.7529541	1.75295414	1.00000000
3-1	8.960000e-01	-0.3014957	2.09349565	0.19209928
4-1	0.000000e+00	-1.7529541	1.75295414	1.00000000
5-1	1.800000e-01	-0.9286655	1.28866555	0.98540645
3-2	8.960000e-01	-0.6718898	2.46388985	0.42122131
4-2	2.220446e-16	-2.0241371	2.02413709	1.00000000
5-2	1.800000e-01	-1.3211402	1.68114024	0.99535491
4-3	-8.960000e-01	-2.4638898	0.67188985	0.42122131
5-3	-7.160000e-01	-1.4999449	0.06794492	0.08103682
5-4	1.800000e-01	-1.3211402	1.68114024	0.99535491

No P-value is smaller than 0.05. The post hoc test used here is not powerful enough to find significant differences among groups.

Tukey's tests should be considered carefully, since they are commonly only appropriate for a balanced plane according to classical linear models [AZA 12].

This is dealt with in section 5.2.3.2. The multifactorial application is tricky since the mathematical formulas vary in relation to the same criteria as those of a multifactorial least-squares ANOVA (see section 5.2.3.2).

5.2.1.2.3. Non-parametric alternative: the Kruskal–Wallis test

This test has no **preliminary applicability conditions**, unlike classical or PERMANOVA, since the values of the quantitative variable (e.g. nitrates) are replaced by their rank. However, it is less powerful than the two previous analyses. It can be carried out with the *kruskal.test()* function, while the Nemenyi post hoc test can be carried out with the *posthoc.kruskal.nemenyi.test()*, which belongs to the [PMCMR] library.

kruskal.test(CP$NO3 ~ as.factor(simprof.phyto))

In this example, the P-value yields a value of 0.02. Therefore, at least one of the groups presents concentrations of nitrates that differ from the others.

library(PMCMR)

posthoc.kruskal.nemenyi.test(x = CP$NO3 , g = as.factor(simprof.phyto), method="Chisquare")

No *P*-value is smaller than 0.05. The post hoc test used here, therefore, is not strong enough to detect significant differences among groups. In any case, examining Figure 5.7 is useful to highlight that the concentrations of nitrates look higher in groups 2 and 3 than in the other groups even if they are not significant.

5.2.2. Quantitative factors that structure gradients

5.2.2.1. Passive (a posteriori) correlations

5.2.2.1.1. Correlations with permutation tests on the axes of a multivariate analysis

These are **correlations of factors to the axes of a multivariate analysis** (unconstrained ordination or multidimensional scaling) carried out on the basis of the permutation tests put forward by Jari Oksanen. The variables used must be quantitative, continuous or discrete. They can be used to determine which external factors (e.g. CP variables) explain the structure observed in the analysis or, in case of non-metric multidimensional scaling, to determine the position of the variables used to obtain the position of the objects (e.g. the biological variables used to determine the structure obtained for the stations).

These correlations are made with the *envfit()* function, which belongs to the [vegan] library. Each variable is correlated independently of the others. Thus, it is possible to test several variables at once to determine those that are significantly correlated to the first two axes and those that are not without any interference among them.

Parametric alternative for unconstrained ordination

These correlations can be applied based on the PCA carried out on the "biology" database. The "chemical–physical" database (CP) and the log-transformed and simplified "phytoplankton" (log1p(phytoS);

variables are the variables tested for correlations. The PCA is carried out as it was before and the variables to be tested are concatenated in the same database ("EF").

PCA_EF<-PCA(scale(bio_simp), scale.unit = TRUE, ncp = 7, graph = FALSE)

EF<-cbind(CP, log1p(phytoS))

acpcor12<-envfit(PCA_EFindcoord[,c(1,2)],EF); acpcor12

	Dim.1	Dim.2	r2	Pr(>r)
Depth	-0.26837	-0.96332	0.2004	0.171
Lumin	0.11947	0.99284	0.1547	0.249
Opt_Depth	**-0.43064**	**-0.90253**	0.4985	**0.006 ****
Temp	0.41212	0.91113	0.1581	0.243
...				
Cryp	-0.24639	0.96917	0.0687	0.591
Eugl	**-0.95104**	0.30908	0.3571	**0.021** *
Phac	-0.50810	0.86130	0.2254	0.126
Stro	-0.15191	**0.98839**	0.3260	**0.033** *
Trac	0.00515	**0.99999**	0.4074	**0.017** *

acpcor13<-envfit(PCA_EFindcoord[,c(1,3)],EF);acpcor13

	Dim.1	Dim.3	r2	Pr(>r)
...				
NO3	-0.27527	**0.96137**	0.3674	**0.018** *
...				
Nitz	**-0.66626**	**0.74572**	0.3862	**0.024** *
...				
Eugl	**-0.98763**	-0.15682	0.3427	**0.049** *

par(mfrow=c(1,2))

plot(PCA_EF,choice="var",axes = c(1,2)); plot(acpcor12,p=0.05)

plot(PCA_EF,choice="var",axes = c(1,3)); plot(acpcor13,p=0.05)

Optical depth, anticorrelated to species diversity on axis 2, and the concentrations of nitrates, anticorrelated to productivity on axis 3 (Figures 5.12(A) and (B)), are the variables that are significantly correlated to the axes for the CP variables. As for phytoplankton taxa, these variables are Euglena, correlated to high abundances and anticorrelated to the Shannon and Simpson diversity on the first axis, *Strombidium* and *Trachelomonas*, correlated to phytoplankton production on the second axis, and *Nitzchia*, correlated to the medium abundances between the first and the third axes (Figures 5.12(A) and (B)).

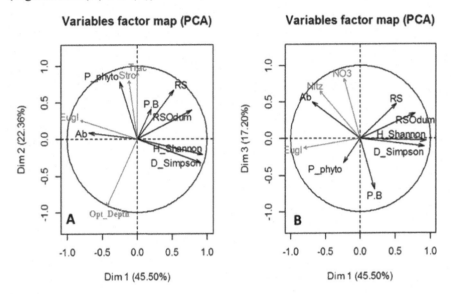

Figure 5.12. *Correlation circle of the second axis versus the first axis (A) and the third axis versus the first one (B) in the principal component analysis carried out on the "global biology" database. The variables that are significantly correlated to the axes of the analysis are shown in gray*

Why should we prefer this analysis to supplementary variables?

Supplementary variables are parameters projected onto the axes of the analysis (especially on the correlation circle for a standardized PCA) **without having been considered in the mathematical calculations related to the analysis** to determine the structure of the objects on the basis of the parameters. Therefore, this is a passive projection that does not affect the results of the analysis. We have already discussed it in section 4.1.2.5. These variables allow us to compare a biological data set with a potentially explanatory data set (i.e. chemical–physical database). Readers should refer to section 4.1.2.5 for its implementation.

Supplementary variables will be dealt with in the same way as active variables, for example with the Euclidean distances for PCA. Thus, supplementary variables must be appropriate for the kind of distance used.

We should not use this type of analysis to compare, for example, a structure based on abundances by means of a CA based on the χ^2 distance with a CP database, for which the Euclidean distance is the most suitable.

Non-parametric alternative for multidimensional scaling

These correlations are applied to the non-metric multidimensional scaling carried out on the "biology" database and interpreted previously in section 4.2.2.2. One of the drawbacks of this analysis was that it did not provide the position of the variables on which the analyses were based, unlike unconstrained ordination. These correlations to the axes based on permutation tests allow us to locate these variables. The correlations corresponding to the biological variables are stored in the **nmdscor** object, while the potentially explanatory variables are stored in **nmdscorexp** (only the significant variables are considered here), so that we can represent them in different colors in the dispersion by type of marsh.

nmdscor<-envfit(NMDSbio,BIO.trf);nmdscor

	NMDS1	NMDS2	r2	Pr(>r)
H_Shannon	-0.99017	-0.13988	0.8385	0.001 ***
RSOdum	-0.91167	0.41092	0.6305	0.003 **
D_Simpson	-0.96039	-0.27866	0.8526	0.001 ***
P_phyto	0.21207	0.97725	0.7044	0.002 **
Ab	0.94219	-0.33508	0.5547	0.009 **
RS	-0.64820	0.76147	0.3291	0.055 .
P.B	-0.12342	0.99235	0.6047	0.001 ***

nmdscorexp<-envfit(NMDSbio,cbind(CP,log1p(phytoS)));nmdscorexp

	NMDS1	NMDS2	r2	Pr(>r)
Depth	0.07443	-0.99723	0.4480	0.007 **
Opt_depth	0.32837	-0.94455	0.4774	0.004 **
Nitz	0.76707	-0.64156	0.3670	0.021 *
Eugl	0.99522	0.09763	0.3557	0.029 *

par(mfrow=c(1,1));s.class(scores(NMDSbio,display="sites"), fac=fext$Type, col=c("dark gray","black"))

plot(nmdscor,p=0.05,col="red"); plot(nmdscorexp,p=0.05,col="blue")

Even if they are not meaningful, some interpretations can be done here for the axes of the nMDS: we thus describe the results using them. All the variables on which the analysis was based are significantly correlated to the axes of the nMDS plane, except for species diversity. The re-fed marsh stations vary more on axis 1, where abundances lie opposite the Shannon and Simpson diversity, whereas the unfed marsh stations vary more on axis 2, which presents a productivity and phytoplankton production gradient (Figure 5.13). The re-fed marshes vary significantly on axis 2 as well. As for potentially explanatory variables, we consider depth and optical depth for CP variables negatively correlated to axis 2 and anticorrelated to productivity and phytoplankton production. As for phytoplankton taxa, *Nitzchia* is anticorrelated to the Odum species diversity and *Euglena* to the Shannon and Simpson diversity (Figure 5.13).

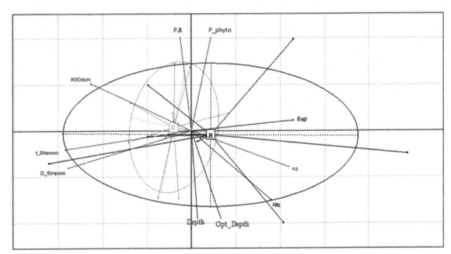

Figure 5.13. *Dispersion of the stations according to the type of marsh (re-fed or not) in the non-metric multidimensional scaling plane realized on the "biology" database. The significant correlated biological variables that have enabled this analysis are shown in red, while the significant correlated chemical–physical and phytoplankton variables are shown in blue. For a color version of this figure, see www.iste.co.uk/ david/data.zip*

5.2.2.1.2. Non-parametric alternative: the Bio-Env procedure

This method tries to find in a potentially explanatory data set (e.g. chemical physics) the best combination of variables that can explain the set of variables that need to be explained (e.g. phytoplankton taxa; [CLA 93]). It is based on Mantel tests that use Spearman's correlation coefficients (based on ranks) between the association matrix based on the data set that has to be explained and different association matrices that consider variables of the explanatory data set. The *bioenv()* function is available in the library [vegan].

A dissimilarity matrix is first calculated from the matrix given in the *comm* argument with the association measure specified in the *index* argument (all the coefficients available in the *vegdist()* function belonging to the library [vegan] can be used). All the potentially explanatory variables specified in the *env* and *upto* arguments are standardized and introduced into a Euclidean distance matrix. The procedure involves looking for the best combination by using first the most explanatory variable in the Euclidean distance matrix, then 2, 3 and so on until variables that no longer improve the correlation are added (Figure 5.14).

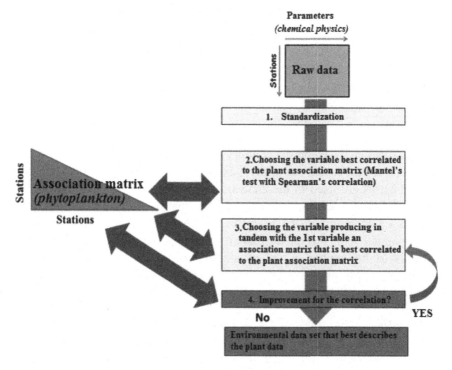

Figure 5.14. *Schematization of the Bio-Env procedure*

This method is used here on the log-transformed (log1p(phytoS)) database that we want to link to the CP variables.

library(vegan)

BIOENV<-bioenv(comm=log1p(bio_simp), CP, method = "spearman", index = "bray",upto = ncol(CP))

BIOENV

Best model has 3 parameters (max. 9 allowed):

Depth PO4 Turb

with correlation 0.4686625

summary(BIOENV) # *It returns the result of all the correlations made*

The best correlation obtained involves depth, the concentrations of phosphates and turbidity, with a final Spearman's correlation coefficient of 0.47.

5.2.2.2. Active regression to the axes of an unconstrained ordination

In the aforementioned methods, the supplementary variables (see section 4.1.2.5) and *a posteriori* passive correlations (see section 5.2.2.1), as well as the structures of elements determined on the basis of active variables (i.e. biological variables in our example) are not affected by potentially explanatory variables. Therefore, these types of analysis are exploratory and descriptive.

Canonical analysis, on the contrary, allows us to explore the relationships between two databases, one of them explained and the other explanatory: the two databases are used in the analysis [LEG 98]. This analysis combines a classical factor analysis/unconstrained ordination (PCA, correspondence analysis or another kind of analysis) and multiple regressions. The factor axes obtained with the unconstrained ordination carried out on the matrix that has to be explained are constrained by multiple regression to the potentially explanatory variables in order to obtain the canonical axes (Figure 5.15).

Therefore, this analysis is strongly influenced by the potentially explanatory variables employed. If these explain badly the variables that have to be explained, which are processed through unconstrained ordination, the structure obtained by ordination runs the risk of being affected too much and the explanatory power of the variables used by regression of being limited. Besides, these types of analysis remain multivariate analyses of a parametric approach and the sum of the explained and explanatory variables must be smaller than the number of elements (i.e. the stations in our case). These two problems can be overcome by carrying out a stepwise regression at the beginning. This type of regression is an automated procedure that allows us to optimize the models of linear regression by choosing only a combination of variables that contribute to the explanation by the relevant variables. This can be done in two ways: (1) with backward selections that choose first the most explanatory variable and then proceed by adding one after the other the variables that provide information, or (2) with backward selection that proceed the other way round by choosing at first all the variables and progressively eliminating those that provide the least amount of information. Neither of these methods is perfect and it is indispensable that researchers consider critically which variables must be taken into account for the model. Bidirectional selection is a combination of both methods ("both") and remains the best compromise. Nonetheless, it is better to carry out a preselection of the potentially explanatory variables beforehand, according to the knowledge provided by the literature, since these approaches are nothing more than mathematical tools.

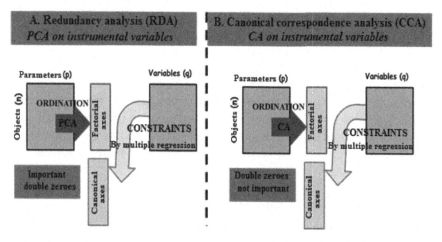

Figure 5.15. *Canonical analysis. (A) A redundancy analysis based at first on a principal component analysis as unconstrained ordination. (B) A canonical correspondence analysis based at first on a correspondence analysis*

If the factor analysis carried out on the variables that must be explained is a type of PCA (based on the Euclidean distance), the corresponding constrained analysis will be called redundancy analysis (RCA) or PCA on experimental variables (Figure 5.15(A)). On the contrary, if the initial analysis is a correspondence analysis (based on the χ^2 distance), the analysis will be a canonical correspondence analysis (CCA) on instrumental variables (Figure 5.15(B)). In other cases (i.e. other association measures), we will refer to constrained principal coordinate analysis or distance-based redundancy analysis (dbRDA). The potentially explanatory variables must be quantitative (continuous or discrete), since they constrain factor analysis through linear models.

Several libraries that allow us to carry out canonical analyses are available in R. The library [vegan] is more complete in the sense that it allows us to carry out permutation tests that yield the significance of the model obtained and the number of significant axes to keep. This can be done with the *cca()* function for CCA (see section 5.2.2.2.1), with the *rda()* function for RDA (see section 5.2.2.2.2), and with the *capscale()* function for general constrained principal coordinate analysis (see section 5.2.2.2.3).

The advantage of these analyses is that they allow us to quantify the variance or inertia explained by potentially explanatory variables. Permutation tests enable us to determine if this explanatory power is significant, as well as the number of axes to keep.

In this kind of analysis, if the potentially explanatory variables belong to different subsets (e.g. CP variables and phytoplankton taxa), a variance partition enables us to determine the part of variance (or inertia) explained by each of the groups considered independently as well as the common part linked to the relationship between these two (or more) subgroups of variables (see section 5.2.2.2.4).

5.2.2.2.1. Canonical correspondence analysis

CCA is used here on the simplified and log-transformed phytoplankton database (15 taxa, "phytoS"; [TER 87a]) on which correspondence factor analysis has already been applied (see section 4.1.3). The CP variables represent the potentially explanatory variables. The number of CP variables to consider to optimize CCA is first determined by applying a bidirectional stepwise regression ("both"). The *cca()* function, which enables us to carry out the CCA belongs to the library [vegan] but can also be found under the same name in the library [ade4] with different arguments. Therefore, we have to start by detaching the library [ade4], since it masks the functions related to the CCA of the library [vegan]. The *cca()* function requires us to specify the database on which the correspondence analysis is carried out, followed by a tilde, a point and the file in which the constraining variables ("CP") are stored, or to specify the relevant variables, separated by +, instead of the period.

library(vegan)

detach("package:ade4", unload=TRUE)

cca <- cca(log1p(phytoS) ~ ., CP) #*The model to consider for stepwise regression, i.e. all the variables included in "CP"*

step.forward<-ordistep(cca(log1p(phyto)   ~   1,   CP),   scope=formula(cca), direction="both", pstep=1000) #*We should look at the last model given*

Step: log1p(phyto) ~ NO3 + NO2

The best model put forward here corresponds to the one that only keeps the concentration of nitrates and nitrites among the set of CP variables. Therefore, the CCA is carried out based on this model and stored in "ccasimp".

ccasimp<-cca(log1p(phytoS)~NO3+NO2, CP) #The model to consider for CCA

R2<-RsquareAdj(ccasimp)$r.squared;R2 #*Calculating the inertia explained by the final model*

[1] 0.2864178

anova.cca(ccasimp,step=1000) # *Significance of the global model*

Model: cca(formula = log1p(phytoS) ~ NO3 + NO2, data = CP)
      Df ChiSquare   F Pr(>F)
Model 2 0.13446 3.211 0.001 ***
Residual 16 0.33501

The final model explains significantly 29% of the total inertia of the phytoplankton database (permutation test of the regressive model, $P = 0.001***$).

anova.cca(ccasimp,by="axis",step=1000) # *Significant axes*

```
Model: cca(formula = log1p(phytoS) ~ NO3 + NO2, data = CP)
         Df ChiSquare    F  Pr(>F)
CCA1      1  0.08244 3.9371 0.001 ***
CCA2      1  0.05203 2.4850 0.010 **
Residual 16  0.33501
```

summary(ccasimp) #*Global view of the results. Here, we only provide the results necessary to calculate the percentage explained by the axes, namely the accumulated constrained eigenvalues*

Accumulated constrained eigenvalues
Importance of components:

	CCA1	CCA2
Eigenvalue	0.08244	0.05203
Proportion Explained	**0.61306**	**0.38694**
Cumulative Proportion	0.61306	1.00000

Only the first two axes are significant (permutation test, $P < 0.05$) and explain $0.61 \times 0.29 = 20\%$ of the inertia for axis 1 and $0.39 \times 0.29 = 11\%$ for axis 2 ("accumulated constrained eigenvalue – proportion explained" $\times$ R2 for each axis).

Two graphical representations can be used: one of them allows us to highlight the relationships between explained ("phytoplankton") and explanatory (CP variables; scaling 1) variables, while the other enables us to analyze more specifically the position of the elements (here, the stations; scaling 2).

plot(ccasimp, scaling=1)

plot(ccasimp, scaling=2)

scores(ccasimp, display="sites")->SC #*Retrieval of the coordinates of the elements (stations)*

library(ade4)

s.class(SC,fac=fext$Station) # *Representation of the dispersion of these elements per marsh*

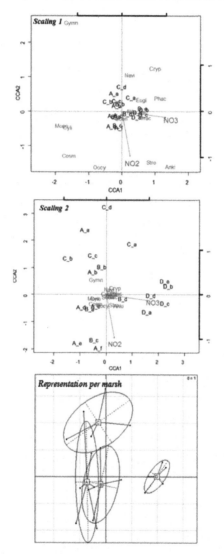

Figure 5.16. *Graphical representation of the canonical correspondence analysis carried out on the simplified and log-transformed phytoplankton database. The representation of the first scaling highlights the relationships between explanatory and explained variables. The representation of the second scaling underlines the structure of the elements. Dispersion of the stations according to the marshes sampled. The axes in bold of the first and second scaling correspond to the correlations of the explanatory variables (arrows)*

On axis 1, *Phacus* (positive values) is opposed to *Cylindrotheca* and *Monoraphidium* (negative values; Figure 5.16). The phytoplankton taxa that are well represented on scaling 1 are *Ooocystis*, *Strombomonas* and *Ankitrodesmus* on the negative values of axis 2, and *Navicula* on the positive values. The concentrations of nitrates are very well linked to axis 1: the higher their values are, the more abundant *Phacus* is (Figure 5.16). The lower the values of these CP parameters are, the higher the abundances of *Cylindrotheca* and *Monaraphidium* are. On axis 2, high concentrations of nitrates correspond to high abundances of *Oocystis*, *Strombomonas* and *Ankitrodesmus*, while low concentrations of nitrites correspond to high abundances of *Cryptomonas* and *Navicula* (Figure 5.16). *Cosmarium* occupies an intermediate position between the two axes, favored by high concentrations of nitrites and low concentrations of nitrates. The stations of marsh D are very markedly distinct from those of the other three marshes, with high concentrations of NO3: they present high abundances of *Phacus*. The stations of marsh C are situated in the positive part of axis 2, namely they show low concentrations of nitrites and lower abundances of *Cryptomonas* and *Navicula*. Finally, the stations of marsh B and marsh A are nearby in the negative values of axis 1 and present high abundances of *Cylindrotheca*, *Monoraphidium* and *Cosmarium* (Figure 5.16).

5.2.2.2.2. Redundancy analysis

RDA is carried on the basis of the PCA performed on the simplified and standardized "biology" database. The explanatory variables used here correspond to the CP and phytoplankton (phytoS) data sets.

library(vegan)

bdd<-cbind(CP,phytoS) #*Grouping the two explanatory databases used*

rda <- rda(scale(bio_simp) ~ ., bdd) #*The model to consider for stepwise regression, i.e. all the variables included in "bdd"*

step.forward<-ordistep(rda(scale(bio_simp) ~ 1, bdd),scope = formula(rda), direction="both",pstep=1000)   #*We have to look at the last model given*

Step: scale(bio_simp) ~ Nitz + Opt_Depth + NO2

The best model put forward here corresponds to the one that only keeps the abundances of *Nitzchia*, optical depth and the concentrations of nitrites among the set of variables considered. Therefore, RDA is carried out based on this model and stored in "rdasimp".

rdasimp<-rda(scale(bio_simp)~Nitz + Opt_Depth + NO2,bdd) *#The model to consider for RDA*

R2<-RsquareAdj(rdasimp)$r.squared;R2 *#Calculating the inertia explained by the final model*

[1] 0.4722248

anova.cca(rdasimp, step=1000) *# Significance of the global model as for CCA*

Model: rda(formula = scale(bio_simp) ~ Nitz + Opt_Depth + NO2, data = bdd)
      Df Variance   F Pr(>F)

Model 3   3.3117 4.4894 0.001 ***

Residual 15   3.6883

The final model explains significantly 47% of the total variance of the "biology" database (permutation test of the regressive model, $P = 0.001***$).

anova.cca(rdasimp,by="axis",step=1000) *# Significant axes as for CCA*

Model: rda(formula = scale(bio_simp) ~ Nitz + Opt_Depth + NO2, data = bdd)
      Df Variance   F Pr(>F)

RDA1   1   2.0084 8.1679 0.008 **

RDA2   1   0.8340 3.3919 0.022 *

RDA3   1   0.4693 1.9084 0.136

Residual 15   3.6883

summary(rdasimp) *#Global view of the results. Here, we only provide the results necessary to calculate the percentage explained by the axes, namely the accumulated constrained eigenvalues as for CCA*

Only the first two axes are significant (permutation test, $P < 0.05$) and explain $0.61 \times 0.47 = 29\%$ of the variance of axis 1 and $0.25 \times 0.47 = 12\%$ of the variance of axis 2 ("accumulated constrained eigenvalue – proportion explained" $\times R^2$ for each axis).

Two graphical representations can be used: one of them allows us to highlight the relationships between the explained ("biology") and the explanatory (CP parameters and phytoplankton; scaling 1) variables, while the other enables us to analyze more specifically the position of the elements (here, the stations; scaling 2).

plot(rdasimp, scaling=1)

plot(rdasimp, scaling=2)

scores(rdasimp,display="sites")->SCrda #*Retrieval of the coordinates of the elements (stations)*

library(ade4)

s.class(SCrda,fac=fext$Station) # *Representation of the dispersion of these elements per marsh*

On axis 1, the Shannon and Simpson diversity and productivity (positive values) lie opposite the total abundances (negative values; Figure 5.17). The *Nitzchia* taxon is well represented among the negative values of this axis: the higher the abundances of this taxon is, the higher the total abundances and the phytoplankton production are. Axis 2 presents the highest taxonomic diversities and phytoplankton productions. These two indices seem well explained by the higher concentrations of nitrites and the low values of optical depth. The stations of unfed marshes (marsh C and marsh D) fluctuate widely on axis 1, namely in relation to the fluctuations of *Nitzchia*: they present the strongest fluctuation in the abundances of this taxon and, therefore, in diversity and productivity. In unfed marshes, the stations of marsh C seem less rich in nitrites and taxa and less productive (positive values on axis 2) than those of marsh D (negative values on axis 2; Figure 5.17).

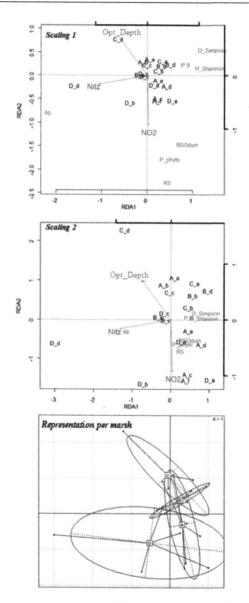

Figure 5.17. *Graphical representation of the redundancy analysis carried out on the standardized and simplified biology database. The representation of the first scaling highlights the relationships between explanatory and explained variables. The representation of the second scaling underlines the structure of the elements. Dispersion of the stations according to the marshes sampled. The axes in bold of the first and second scaling correspond to the correlations of the explanatory variables (arrows)*

5.2.2.2.3. Constrained principal coordinate analysis or dbRDA

This analysis represents a generalization of constrained factor analysis. In this sense, it allows us to carry out at the beginning **factor analysis with the distance or dissimilarity of our choice**. Thus, the applicability conditions depend on the distance/dissimilarity chosen.

Constrained PCA is carried out here on the log-transformed and simplified phytoplankton database on which CCA has been used (see section 5.2.2.2.1) but using Bray-Curtis dissimilarities. The CP variables (see section I.4.2) are the potentially explanatory variables used here. The results are slightly less detailed than those obtained with CCA, since they are fairly similar. The same explanatory variables are kept through stepwise regression.

library(vegan)

dbRDA <- capscale(log1p(phytoS) ~ ., CP, dist="bray")

step.forward<-ordistep(capscale(log1p(phyto) ~ 1, CP),scope=formula(dbRDA),

   direction="both",pstep=1000)

Step: log1p(phyto) ~ NO3 + NO2

dbRDAsimp<- capscale(log1p(phytoS)~ NO3+NO2 , CP, dist="bray")

R2<-RsquareAdj(dbRDAsimp)$r.squared;R2

[1] 0.2865351

anova.cca(dbRDAsimp,step=1000) #As for CCA and RDA

anova.cca(dbRDAsimp,by="axis",step=1000)# As for CCA and RDA

summary(dbRDAsimp) *#As for CCA and RDA, we have to consider the proportion explained in the part accumulated constrained eigenvalues*

dbRDA explains significantly 29% of the inertia of phytoplankton with two axes ($0.60 \times 0.29 = 17\%$ for axis 1 and $0.40 \times 0.29 = 12\%$ for axis 2).

plot(dbRDAsimp, scaling=1)

plot(dbRDAsimp, scaling=2)

scores(dbRDAsimp,display="sites")->SCdbrda

library(ade4); s.class(SCdbrda,fac=fext$Station)

Cosmarium, Monoraphidium and *Cylindrotheca* taxa are well represented among the positive values of axis 1 and anticorrelated to nitrates. Therefore, they are more abundant when the concentrations of nitrates are low. *Strombomonas* and *Ankitrodesmus* taxa are positively linked to axis 2, as are the nitrites. Thus, they are more abundant in relation to high nitrites concentrations, unlike the *Gymnodinium* taxon (Figure 5.18).

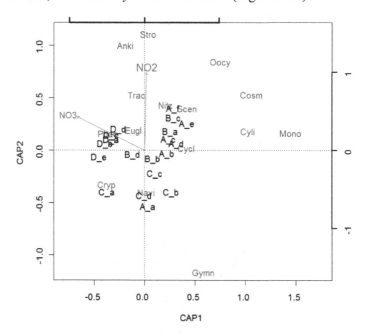

Figure 5.18. *Projection of the dbRDA carried out on the simplified phytoplankton database constrained by the chemical–physical variables (scaling 1)*

5.2.2.2.4. Variance partitioning

In a canonical analysis, we can **estimate the variance (or inertia) associated with subsets of variables**. For example, when we carry out RDA (see section 5.2.2.2.2), the explanatory variables corresponded to two types of variables: CP variables (optical depth and concentrations of nitrates) and phytoplankton taxa (*Nitzchia*). Variance partitioning allows us to quantify:

1) the pure part in the explanation of the structure of the stations based on the biological variables explained by the CP parameters;

2) the pure part explained by the phytoplankton taxa;

3) the mixed part between the two sets of variables linked to the relationship between the two data sets.

Variance partitioning can be performed for several submatrices, even if in this case we only use two of them. It can be carried out with the **varpart()** function belonging to the library [vegan]. It requires us to specify the initial database and then the two potentially explanatory matrices.

PHYS<-CP[,c("Opt_Depth","NO3")]

PHYTO<-phytoS[,c("Nitz")]

library(vegan)

varpart(scale(bio_simp),PHYS,PHYTO)->VP

VP

plot(VP)

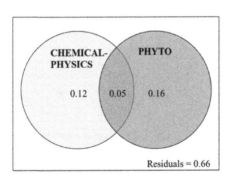

Figure 5.19. *Variance partitioning carried out on the redundancy analysis performed on the global biological database. The potentially explanatory variables have been classed into two subgroups: "chemical–physics" and "phytoplankton"*

The CP base explains purely 12% of the variance of the biology base, while the phytoplankton base explains 16% of it. A mixed part between the two explains 5% of it. Sixty-six percent remains unexplained by these two data matrices (Figure 5.19).

5.2.3. Qualitative factors that structure gradients

5.2.3.1. Parametric alternative: MANOVA and discriminant function analysis

A MANOVA-like test must be carried out beforehand in order to determine whether the modalities of the qualitative variable can be identified in a significant way on the basis of the quantitative variables considered (section 5.2.3.1.1) before performing discriminant function analysis (DFA; sections 5.2.3.1.2 and 5.2.3.1.3).

5.2.3.1.1. Multivariate analysis of variance

MANOVA [HAN 87] is a **statistical test that allows us to determine the effect of one or more qualitative variables in a matrix of quantitative variables**. It represents an ANOVA extrapolated to multivariate analysis. It constitutes a parametric test that can be applied to data compatible with the Euclidean distance, i.e. double zeroes are significant. The applicability conditions are the following: homogeneity of the variances, independence of the objects and normality of the data. The alternative hypothesis assumes that at least one of the modalities of the qualitative factor responds differently to the quantitative variables. There are several tests: Pillai's trace is the most robust in relation to the applicability conditions and the normality of the data in particular.

Here, MANOVA is carried out on the standardized matrix of CP data (see section I.4.2) to test one by one the qualitative variables available in the "environmental factors" database ("fext"). The *manova()* function is available in the base version of R. The *summary()* function allows us to specify the test chosen, which in this case is Pillai's trace.

```
m1 <- manova(as.matrix(scale(CP)) ~ fext$Station)

summary(m1, test = "Pillai") #P-value=*

m1 <- manova(as.matrix(scale(CP)) ~ fext$Type)
```

summary(m1, test = " Pillai ") #*P-value=*

m1 <- manova(as.matrix(scale(CP)) ~ fext$Position)

summary(m1, test = " Pillai ") #*P-value=ns*

Only the "Marsh" and "Type" variables present a result for which the null hypothesis is rejected (*P*-value < 0.05, MANOVA and Pillai's trace test): at least one of the marshes differs from the other based on at least one of the CP parameters. Thus, the two types of marsh (re-fed or not) differ on the basis of CP parameters.

5.2.3.1.2. Classical DFA

DFA [TER 87b] is a type of **constrained factor analysis that aims to determine the quantitative variables that allow us to separate as well as possible the point clouds corresponding to the modalities of a qualitative variable**. It uses optimization based on the principle of a PCA to look for axes (at most the variable number –1) on which the projections of the clouds are separated as well as possible (Figure 5.20). Being based on the same principle as a PCA, it is subjected to the same applicability conditions; in particular, double zeroes must be considered as a factor of similarity between objects (see section 4.2.2.1). It can be carried out with the *discrimin()* function, among others, which belongs to the library [ade4], by specifying the result of a PCA performed with the library [ade4]: the *dudi.pca()* function.

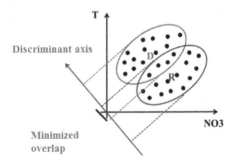

Figure 5.20. *Principle of discriminant function analysis: example of the optimization of an axis that best discriminates the two point clouds (re-fed or unfed marshes) in the plane formed by these two variables: temperature and concentrations of nitrates*

library(ade4)

acp<-dudi.pca(scale(CP), nf=ncol(CP),scan = F)

dis <- discrimin(acp, fext$Station,nf=ncol(CP),scan = F)

dis

Discriminant analysis
call: discrimin(dudi = acp, fac = fext$Station, scannf = F, nf = ncol(CP))
class: discrimin

$nf (axis saved) : 3

eigen values: 0.9864 0.8308 0.2054

	data.frame	nrow	ncol	content
1	$fa	9	3	loadings / canonical weights
2	$li	19	3	canonical scores
3	$va	9	3	cos(variables, canonical scores)
4	$cp	9	3	cos(components, canonical scores)
5	$gc	4	3	class scores

The analysis has kept three discriminant axes. The dis$fa table provides the canonical weights that yield the linear combinations of the variables on the basis of which we can calculate the contributions of each variable to the axes as follows.

FDA1= abs(dis$fa[,1])/sum(abs(dis$fa[,1]))*100

FDA2= abs(dis$fa[,2])/sum(abs(dis$fa[,2]))*100

FDA3= abs(dis$fa[,3])/sum(abs(dis$fa[,3]))*100

contribAFD=data.frame(var=colnames(CP),FDA1,FDA2, FDA3)

round(contribAFD,0)

var	FDA1	FDA2	FDA3
Depth	1.8353968	21.6225328	18.618586
Lumin	12.6413499	4.0648815	11.300844
Opt_Depth	5.1068102	0.3418222	6.265403
Temp	1.5823951	30.9653640	8.232386
NO2	12.6645940	2.0569486	11.790422
NO3	60.5600935	19.1480632	2.021121
PO4	0.3544859	18.4547992	6.614903
N.P	0.9757287	2.8265114	6.534316
Turb	4.2791459	0.5190770	28.622018

par(mfrow=c(2,2))

s.class(dis$li,fac=fext$Station,xax=1,yax=2,col=c(1:4))

    s.label(dis$fa)

s.class(dis$li,fac=fext$Station,xax=1,yax=3,col=c(1:4))

    s.label(dis$fa[,c(1,3)])

The first two axes discriminate on their own four marshes (Figure 5.21(a)): only marsh A and marsh B overlap slightly. Axis 3 does not contribute anything as for the discrimination of these two marshes (Figure 5.21(c)). The concentrations of nitrates contribute the most to the construction of the first axis (61%). Temperature (31%), depth (22%), and the concentrations of nitrates (19%) and phosphates (18%) contribute essentially to the construction of the second axis.

Therefore, the explanatory model can be simplified by only keeping those variables that contribute essentially to the construction of these two axes: the concentrations of nitrates and phosphates, depth and temperature. The dispersion of the stations per marsh is also as good as it is for the model that incorporates all the variables (Figure 5.22).

Afterward, it is possible to use the model obtained to predict to which modality the qualitative variable (here, the marsh) belongs based on the data of the variables obtained (i.e. CP variables).

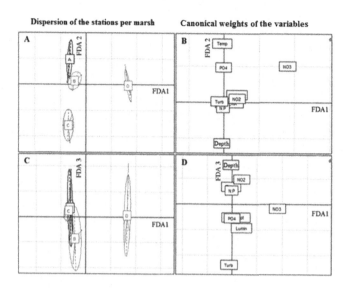

Figure 5.21. *Discriminant function analysis carried out to determine which chemical–physical variables allow us to discriminate the marshes: dispersion of the stations according to the second discriminant axis in relation to the first discriminant axis (A) and according to the third discriminant axis in relation to first discriminant axis (C). Canonical weights of the variables on the second discriminant axis in relation to the first discriminant axis (B), and on the third discriminant axis in relation to the first discriminant axis (D)*

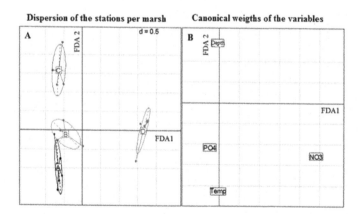

Figure 5.22. *Discriminant function analysis carried out to determine which chemical–physical variables allow us to discriminate the marshes by only keeping the four most discriminant variables: dispersion of the stations according to the second discriminant axis in relation to the first discriminant axis (A). Canonical weights of the variables on the second discriminant axis in relation to the first discriminant axis (B)*

5.2.3.1.3. Discriminant correspondence analysis

Discriminant correspondence analysis is identical to DFA and based on the same mathematical principle, except for the fact that it relies on a correspondence analysis rather than on a PCA. Therefore, it is a type of constrained analysis that aims to determine the quantitative variables for which double zeroes are not a determining criterion in the assessment of the similarities between objects. It searches by optimization new axes (at most the number variable −1) for which the projections of the clouds are separated as well as possible based on a CA (Figure 5.20). Thus, it is subjected to the same applicability conditions as correspondence analyses (see section 4.1.3.1). It can be carried out with the ***discrimin.coa()*** function available in the library [ade4]. It will be performed here based on the simplified and log-transformed "phytoplankton" database (phytotrf, see section 4.2.3.2).

The analysis has kept two discriminant axes that isolate the stations of the four marshes (Figure 5.23(A)). *Cosmarium* (15%), *Strombomonas* (13%), *Gymnodinium* (12%), *Scenedesmus* (11%) and *Cryptomonas* (10%) taxa contribute essentially to the construction of discriminant axis 1. *Cylindrotheca* (19%), *Cosmarium* (15%), *Trachelomonas* (11%) and *Gymnodinium* (10%) taxa contribute essentially to the construction of axis 2. The stations of marsh C are influenced by the *Cosmarium* and *Gymnodinium* taxa, the stations of marsh D are influenced by the *Cryptomonas*, *Strombomonas* and *Scenedesmus* taxa, while those of marsh A by the *Cylindrotheca* and *Trachelomonas* taxa (Figures 5.23(a) and (b)).

```
library (ade4)

afcD<-discrimin.coa(phytotrf, fext$Station, scan = FALSE)

summary(afcD)

afcD

FDA1= abs(afcD$fa[,1])/sum(abs(afcD$fa[,1]))*100

FDA2= abs(afcD$fa[,2])/sum(abs(afcD$fa[,2]))*100

contribafcD=data.frame(var=colnames(phytotrf),FDA1,FDA2)

contribafcD
```

var	FDA1	FDA2
Cycl	5.1371146	3.3378054
Navi	1.5189563	5.6093069
Cyli	2.3851678	19.3920006
Nitz	2.6161029	8.9883869
Gymn	11.4743426	10.3446181
Cosm	14.6845000	14.6845791
Anki	1.0602695	2.8365668
Mono	6.1492790	2.8879641
Oocy	6.5896008	2.9530807
Scen	11.4333437	2.9196000
Cryp	10.9641096	7.7611036
Eugl	0.4848484	4.0265925
Phac	6.2152347	2.6415637
Stro	13.4614561	0.7813125
Trac	5.8256739	10.8355192

par(mfrow=c(1,2))

s.class(afcD$li,fac=fext$Station,xax=1,yax=2,col=c(1:4));s.label(afcD$fa)

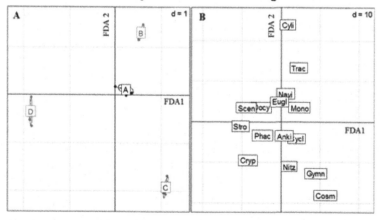

Figure 5.23. *Discriminant correspondence analysis carried out to determine which phytoplankton taxa allow us to discriminate the marshes: dispersion of the stations according to the second discriminant axis in relation to the first discriminant axis (A). Canonical weights of the variables on the second discriminant axis in relation to the first discriminant axis (B)*

5.2.3.2. Intermediate alternative: permutational MANOVA or PERMANOVA

This analysis allows us to test the **simultaneous response of one or more variables to one or more qualitative factors** [AND 01]. This type of analysis is being used more and more in ecology due to:

– **its robustness**: this analysis is not based on any hypothesis on the distribution of the variables. It can even be applied to qualitative variables, provided that they are transformed into binary variables with a complete disjunctive table (see section 2.2.3) and we choose a suitable association measure (see section 2.3.1). However, this analysis needs to test the homogeneity of the multivariate dispersions among groups. It is much more robust than the analysis of similarities and the Mantel test in relation to the heterogeneity of the dispersions [AND 13]. Moreover, it is sensitive, just like these two tests, to unbalanced planes ;

– **its flexibility**: this analysis is based on association matrices. This makes it more powerful than MANOVA (see section 5.2.3.1), since the association measure can be chosen in relation to the goals and the composition of the database.

We have already described the principle of PERMANOVA in relation to a unifactorial approach (section 5.2.1.2.2). It can be applied to a wide range of experimental planes or complex sampling, just like parametric ANOVAs. Moreover, unlike the other non-parametric approaches, it allows us to test the effect of the interactions between factors and to approach variance partitioning.

The PERMANOVA is implemented in R in the library [vegan] with the *adonis()* function. The betadisper test, which allows us to test the homogeneity of the dispersions, can be carried out with the *betadisper()* function belonging to the library [vegan]. It requires us to enter the transformed database, a tilde and the factor(s) considered separated by * in order to specify the interaction calculations, the database in which the factors are presented in the *data* argument, the association measure chosen in the *method* argument and, finally, the permutation number used to calculate the theoretical distribution in the *perm* argument. All the coefficients available in the *vegdist()* function of the library [vegan] can be used.

This analysis has been previously used on a simple plane (see section 5.2.1.2.2). Here, it is carried out for phytoplankton abundances. In this case, we have chosen the Bray–Curtis dissimilarity as association measure, which is the most suitable for the data. Marshes and position (internal and external) represent the explanatory qualitative factors chosen. They are crossed factors.

perm<-adonis(phytotrf ~ Marsh*Position, data=fext, method = "bray", permutations=9999)
perm

Permutation: free
Number of permutations: 9999

Terms added sequentially (first to last)

	Df	Sumsofsqs	MeanSqs	F.Model	R2	Pr(>F)
Station	3	0.62667	0.208889	2.5931	0.34151	0.0015 **
Residuals	15	1.20832	0.080554		0.65849	
Total	18	1.83498			1.00000	

Signif. codes: 0 '***' 0.001 '**' 0.01 '*' 0.05 '.' 0.1 ' ' 1

The type of marsh (drained or re-fed) explains significantly 17% of the dissimilarities between the stations on the basis of the phytoplankton communities (P-value = 0.003; PERMANOVA). Even if the position of the marshes (internal or external) does not seem a significant factor (P-value = 0.386), the interaction between type and position explains significantly 12% of the dissimilarities between stations on the basis of the phytoplankton communities (P-value = 0.013): the dissimilarities between types of marshes are not the same according to the position of the marshes. For each significant factor, the PERMANOVA can be carried out once again in order to understand the differences between the two types of factors (see section 5.2.1.2.2).

The different planes (i.e. crossed factors versus hierarchized factors, mixed models versus fixed models) are incremented like in ANOVA. Refer to the help section of the function for its application.

Conclusion

Always keep in mind the goals

It is very easy to fall into the trap of numerical analysis once we have dived in: one analysis leads to another, then another and so on. Readers should regularly recall their objectives by wondering whether the analysis being carried out provides any information in relation to the objectives of the study. The same data set allows us to achieve different goals. If, for example, our analysis aims to highlight a structure of the stations on the basis of biological communities and to underline the relationships between this structure and the environment, a canonical analysis is more suitable than a principal component analysis on the environmental variables (even if the latter analysis may be carried out for certain reasons, e.g. to explore some relationships and, therefore, certain redundancies between environmental variables).

In theory, the type of numerical analysis should have already been defined when the sampling strategy or experimental plan conceived is implemented in order to avoid unpleasant surprises at the end, i.e. lack of data for a more powerful parametric analysis, unsuitable sampling or experimental plan, etc.

Finally, it is very important to specify clearly in our material and method, for all kinds of analysis, the consistency between the numerical analysis carried out and the goal, so that the reader can get a good grasp of the consistency between the scientific and the statistical process.

Rigorousness in the analyses

Numerical analyses must be used rigorously. We have to be aware that the mathematical principles on which these analyses are based are strict. If they are not respected, the analyses will be biased and, in this case, we can actually interpret data in any possible way... For example, if a small number of objects compared to the number of variables are available, parametric approaches should be avoided. Similarly, an association measure must be chosen in relation to the data set and our objectives (i.e. double zeroes).

Enhancing data without exaggerating

Naturally, numerical analyses allow us to enhance data and to sum up the information of heavy data sets in a few graphs. However, excessive data manipulation is another trap we should avoid if we do not want to hear that we are interpreting data our own way, whereas the goal of the analyses is to remain as objective as possible in relation to a large multivariate data set.

Several analyses enable us to achieve the same goal, even if a data set already directs us toward certain kinds of analysis. Nonetheless, several options are available: several classification algorithms, several types of multivariate analysis, etc. For some of these, there are analyses that allow us to account for the analysis that is most suitable for the data set (e.g. cophenetic matrix for classification algorithms). For others, we will have to try several types of analysis and choose the **most informative** (rather than convenient!). It is useless to present several analyses that provide the same information. This only multiplies the number of analyses and confuses the reader. If we are led to choose several analyses, it is essential that they complement each other. Finally, if we have to choose between two redundant analyses or approaches, we should favor the **simpler** one.

Some types of analysis need not be illustrated in the results

Some analyses are only presented in the material and method, since they correspond to a preliminary stage, such as the elimination of rare species. However, it is important to specify these analyses as well as their results in our material and method.

Bibliography

[AND 01] ANDERSON M.J., "A new method for non-parametric multivariate analysis of variance", *Austral Ecology*, vol. 26, pp. 32–46, 2001.

[AND 13] ANDERSON M.J., WALSH D.C.I., "PERMANOVA, ANOSIM, and the Mantel test in the face of heterogeneous dispersions: What null hypothesis are you testing?", *Ecological Monographs*, vol. 83, no. 4, pp. 557–574, 2013.

[AZA 12] AZAÏS J.-M., BARDET J.-M., *Le modèle linéaire par l'exemple*, Dunod, Paris, 2012.

[BOR 11] BORCARD D., GILLET F., LEGENDRE P., *Numerical Ecology with R*, Springer, New York, 2011.

[BRA 57] BRAY J.R., CURTIS J.T., "An ordination of upland forest communities of southern Wisconsin", *Ecological Monographs*, vol. 27, pp. 325–349, 1957.

[BRE 84] BREIMAN L., FRIEDMAN J.H., OLSHEN R.A. *et al.*, *Classification and Regression Trees*, Wadsworth & Brooks/Cole Advanced Books & Software, Monterey, 1984.

[CHA 08] CHAPMAN M.G., UNDERWOOD A.J., "Scales of variation of gastropod densities over multiple spatial scales: comparison of common and rare species", *Marine Ecology Progress Series*, vol. 354, pp. 147–160, 2008.

[CLA 93a] CLARKE K.R., "Non-parametric multivariate analyses of changes in community structure", *Australian Journal of Ecology*, vol. 18, pp. 118–127, 1993.

[CLA 93b] CLARKE K.R, AINSWORTH M., "A method of linking multivariate community structure to environmental variables", *Marine Ecology Progress Series*, vol. 92, pp. 205–219, 1993.

[CLA 08] CLARKE K.R., SOMERFIELD P.J., GORLEY R.N., "Testing null hypotheses in exploratory community analyses: similarity profiles and biota-environmental linkage", *Journal of Experimental Marine Biology and Ecology*, vol. 366, pp. 56–69, 2008.

[DUF 97] DUFRÊNE M., LEGENDRE P., "Species assemblages and indicator species: the need for a flexible asymmetrical approach", *Ecological Monographs*, vol. 67, pp. 345–366, 1997.

[ESC 70] ESCOUFIER Y., "Echantillonnage dans une population de variables aléatoires réelles", *Publication de l'Institut Statistique de l'Université de Paris*, vol. 19, pp. 1–47, 1970.

[ESC 94] ESCOFIER B., PAGES J., "Multiple factor analysis (ALFMUT package)", *Computational Statistics and Data Analysis*, vol. 18, pp. 121–140, 1994.

[GAU 14] GAUVRIT N., *Statistiques: méfiez-vous!*, Ellipses Marketing, Paris, 2014.

[GOW 75] GOWER J.C., "Generalized Procrustes analysis", *Psychometrika*, vol. 40, pp. 33–50, 1975.

[GRA 05] GRALL J., COIC N., Synthèse des méthodes d'évaluation de la qualité du benthos en milieu côtier, Report, Directive Cadre Stratégie Milieu Marin, 2005.

[GRO 14] GROSJEAN P., IBAÑEZ F., "PASTECS–Package for Analysis of Space-Time Ecological Series", user manual, available at http://www.sciviews.org/pastecs, 2014.

[HAN 87] HAND D.J., TAYLOR C.C., *Multivariate Analysis of Variance and Repeated Measures*, Chapman and Hall, 1987.

[HOT 33] HOTELLING H., "Analysis of a complex of statistical variables into principal components", *Journal of Educational Psychology*, vol. 24, pp. 417–441, 1933.

[IBA 93] IBAÑEZ F., DAUVIN J.C., ETIENNE M., "Comparaison des évolutions à long terme (1977-1990) de deux peuplements macrobenthiques de la baie de Morlaix (Manche occidentale): relations avec les facteurs hydroclimatiques", *Journal of Experimental Biology and Ecology*, vol. 169, pp. 181–214, 1993.

[JAI 88] JAIN A.K., DUBES R.C., *Algorithms for Clustering Data*, Prentice Hall, Englewood Cliffs, 1988.

[KRU 78] KRUSKAL J.B., WISH M., *Multidimensional Scaling*, Sage Publications, Beverly Hills, 1978.

[LEG 98] LEGENDRE P., LEGENDRE L., *Numerical Ecology*, 2nd ed., Elsevier, Amsterdam, 1998.

[LUK 51] LUKASZEWICZ J., "Sur la liaison et la division des points d'un ensemble fini", *Colloq. Math.*, vol. 2, pp. 282–285, 1951.

[MAN 67] MANTEL N., "The detection of disease clustering and a generalized regression approach", *Cancer Research*, vol. 27, pp. 209–220, 1967.

[MCA 01] MCARDLE B.H., ANDERSON M.J., "Fitting multivariate models to community data: a comment on distance-based redundancy analysis", *Ecology*, vol. 82, pp. 290–297, 2001.

[ROU 67] ROUX G., ROUX M., "À propos de quelques méthodes de classification en phytosociologie", *Review of Statistics and its Application*, vol. 15, pp. 59–72, 1967.

[SCH 84] SCHERRER B., *Biostatistique*, Gaetan Morin, Quebec, 1984.

[SNE 73] SNEATH P.H.A., SOKAL R.R., *Numerical Taxonomy – The Principles and Practice of Numerical Classification*, W.H. Freeman, San Francisco, 1973.

[SØR 48] SØRENSEN T., "A method of establishing groups of equal amplitude in plant sociology based on similarity of species content and its application to analysis of the vegetation on Danish commons", *Biologiske Skrifter*, vol. 5, pp. 1–34, 1948.

[TER 87a] TER BRAAK C.J.F., "The analysis of vegetation-environment relationships by canonical correspondence analysis", *Vegetatio*, vol. 69, nos. 69–77, pp. 594–600, 1987.

[TER 87b] TER BRAAK C.J.F., "Calibration", in JONGMAN R.H.G., TER BRAAK C.J.F., VAN TONGEREN O.F.R. (eds.), *Data Analysis in Community and Landscape Ecology*, Pudoc, Wageningen, 1987.

[TOR 58] TORGERSON W.S., *Theory and Methods of Scaling*, John Wiley, New York, 1958.

[UMA 10] UMAÑA-VILLALOBOS G., "Temporal variation of phytoplankton in a small tropical crater lake, Costa Rica", *Revista de Biología Tropical*, vol. 58, pp. 1405–1419, 2010.

[WAR 63] WARD J.H., "Hierarchical grouping to optimize an objective function", *Journal American Statistical Association*, vol. 58, pp. 236–244, 1963.

Index

Printed in the United States
By Bookmasters